T0361939

ROUTLEDGE LIBRARY EDITIONS: AGRICULTURE

Volume 14

OLD FARMS AND NEW FARMING

OLD FARMS AND NEW FARMING

A Layman's View of the Land

CHARLES FURTH

Routledge
Taylor & Francis Group
LONDON AND NEW YORK

First published in 1975 by George Allen & Unwin Ltd

This edition first published in 2020
by Routledge
2 Park Square, Milton Park, Abingdon, Oxon OX14 4RN

and by Routledge
52 Vanderbilt Avenue, New York, NY 10017

Routledge is an imprint of the Taylor & Francis Group, an informa business

British Library Cataloguing in Publication Data
A catalogue record for this book is available from the British Library

ISBN: 978-0-367-24917-5 (Set)
ISBN: 978-0-429-32954-8 (Set) (ebk)
ISBN: 978-0-367-25113-0 (Volume 14) (hbk)
ISBN: 978-0-429-28616-2 (Volume 14) (ebk)

Publisher's Note
The publisher has gone to great lengths to ensure the quality of this reprint but points out that some imperfections in the original copies may be apparent.

Disclaimer
The publisher has made every effort to trace copyright holders and would welcome correspondence from those they have been unable to trace.

OLD FARMS AND NEW FARMING

A Layman's View of the Land

by

Charles Furth

Speak to the earth and it shall teach thee
Job XII, 8

London George Allen & Unwin Ltd
Ruskin House Museum Street

First published in 1975

ISBN 0 04 630008 2

Printed in Great Britain
in 13 point Bembo type
by Acolortone Ltd
Ipswich, Suffolk

To

STRUTT AND PARKER (FARMS) LTD

whose farming I see from my doorstep

and to my neighbour

WILLIAM FAIERS

from whose quiet wisdom I benefit as they did

Preface

To my mind, if you have never seen the harvest brought home, you have yet to see one of the most moving sights you will ever see. For this means security of life and livestock through another winter. And the excitement comes up from our most primitive roots.

We do not live on oil or plastics, nor on advertising and deodorants, although the first two have become essential to us. What we live on is grain, roots, grass and the livestock which shares them with us, and can eat the grass directly. Although most countries are over-urbanised, we are becoming increasingly aware again that our immediate concern is what we are going to eat and what it costs.

Farming in developed countries is mechanised, sophisticated, precise – yet it still depends on season and sun, rain and frost (at the right time and in the right amounts) and on a man's judgement of the state of the soil. I write for those with a new interest in our basic industry, for those who increasingly visit, retire to or commute from the country and want to understand and assess the operations going on in the fields and barns around them. To the knowledgeable my dedication reveals my strengths and my limitations, but I write for all those who take pride in the efficiency of British agriculture, unrivalled in most of Western Europe.

Acknowledgements

Thanks are due to the following for the use of illustrations: Agricultural Press Ltd No. 12; *The Farmer's Weekly* Nos 16 and 18 (photos Peter Adams); Farmhand (U.K.) Ltd No. 19; Imperial Chemical Industries No 13; Mr V. Lakey colour plates Nos I and II; Mr Lakey and the *East Anglian Magazine* No. 1; Massey Ferguson Nos 2, 3, 8, 9 and 20; Mike Randall Photography Ltd No. 7; the Milk Marketing Board No. 11; *Pig Farming* No. 17; Ransomes, Sims and Jeffries Nos 4, 5 and 6; the University of Newcastle Department of Photography Nos 10 and 15.

I must also acknowledge the help of that encyclopedic book of reference, Fream's *Elements of Agriculture* in its latest edition edited by Dr D. H. Robinson (John Murray).

To Mr J. N. Merridew, Director of the University of Newcastle's Nafferton Farm, I owe thanks for his help with illustrations and for what I have learned from his own short textbook, *Farming*, (EUP).

Contents

1 *Landscape with potatoes*

Illustrations

Note on Names

Some farm animals, like some members of the upper British nobility, go by different names at different periods of their lives, and this can lead to confusion.

Cattle
A *heifer* is called a cow after producing her first calf.
A *bullock* or *steer* is a castrated bull calf.

Sheep
A *hogget* is a sheep of either sex between one and two years old.
A *wether* is a castrated ram.
A *tup* is a dialect term for a ram.

Pigs
A *gilt* becomes a sow after producing her first litter.

Poultry
A *pullet* becomes a hen after completing her first egg-laying cycle, at her first moult.

I have deliberately used the names of the old farming counties, rather than those of the new and larger local government areas, which could not convey the same meaning.

WHEAT BARLEY OATS

1
THE ARABLE YEAR

The stubble field lies under the autumn sun, bare as a hog's back except for the stiff bristles of the corn stubble. Their pale gold is reflected down the valley in the solid golden lump of the baled straw, packed high to the roof of the open-sided Dutch barn. The grain harvest is in; the straw all baled and carted; the field empty and silent except for the twittering groups of sparrows.

This, the end of the harvest, is the beginning of the arable year, for the soil is never idle. From now until next February the plough and the harrow will be going up and down, burying the residues of last year's crop, preparing the soil for the new. Whenever the land can safely take the tractor's weight its steady chatter will float on the wind, eight to ten hours a day. The indoor worker, returning home in the winter dusk will be startled by the two surprisingly close-set headlamps coming towards him on no road, with the rear spotlight still watching the movement of the plough.

For the winter-sown barley (which generally goes in first) or winter wheat or oats, it may be enough just to harrow the stubble field before drilling in the new seed. But this is not the

2 *A disc harrow*

light harrow used in the spring to break up the weathered clods of the ploughed field. This is a powerful instrument whose massive prongs or tines pulverise the soil as they are dragged through it, and disc harrows (strange rows of metal discs, which look like heavy cymbals strung along an axle and running on their edges behind the tractor) may break down the earth still further. But the seed bed for an autumn-sown cereal should be firm – some fine, crumby material, yes, but also a good number of clods the size of your fist, for they prevent the soil being 'panned' into a flat, crusted pancake by heavy winter rain and they also give some protection to young seedlings against dry and frosty winds.

The tractor driver does not sit solid in the cabin of his tractor like the lorry driver in his cab. Rather he rides the tractor almost like a horse, because it moves across the field with something of the gait of a cantering horse. The movement on the whole of course is steady, but watch the upright exhaust-pipe closely against the background and you see that the tractor progresses in jerks, is momentarily held by the resistance of earth and stubble roots to the tearing of the tines, overcomes them and bounds on; not the steady movement of a car on a metalled road but rather that of a small ship against the waves. And the man on the iron seat is never still, constantly watching the cultivator or harrow behind him, constantly moving his steering-wheel, hour after hour.

Sometimes this autumn cultivating is to clean the stubbles after a previous crop before the field is ploughed, and the plough is still the usual instrument for soil which is to be weathered through the winter, before the seed is drilled in the spring. This is man's immemorial tool, in its modern form mounted on a massive steel beam, with a coulter, or sharp steel disc revolving in front to make the vertical cut in the soil; the share, or nose of the frame in front of the mould-board to make the horizontal cut, and the mould-board itself, which may be three feet or more in length, to turn the furrow on its side. But where the ploughman following the plough behind his horse team was turning a single furrow, today's plough cuts two, three, four or six furrows at once and – a point to emphasise – so quick has been the change, that the man ploughing the

3 *A three-furrow, reversible plough. The disc coulters and shining mould-boards not in use show up well. On the return journey they will take the place of the three now in the soil*

field may well be the same man who once ploughed it with horses.

The single-furrow plough was more complicated to work, for you do not want your furrows turned in alternate directions, as you work to and fro. Consequently you went up one half of the field, across, and down the other. Now your set of ploughs is duplicated: as you reach the end of one group of furrows you change to your second set of ploughs whose mould-boards, as you go back down the field, turn the furrows in the same direction as those turned by the first set on the way up. Thus you can work steadily up and down the field. Ploughing, even by machine, is still an art. The depth must be even; the land left as even as possible – the ridges not too high nor the furrows too deep; all rubbish and manure must be well buried: bad ploughing cannot be rectified later and will reduce the crop because of bad germination. This is a precision job: the disc coulter should be set three- to five-eighths of an inch out of straight, so that the side towards the unploughed land is not pressed too hard against the furrow wall. (As with all cutting tools, there must be a clearance behind the cutting edge.)

It is an art too, learned only by experience, to know when to get on with the job which so urgently wants to be done, and when giving way to impatience does more harm than good and leaves the soil a morass. I remember – and the weather was dangerously wet – an immense four-wheel-drive tractor (speaking to the driver my head was at his knee) drawing a four-furrow plough. The wet, heavy soil stuck to the tractor wheels, the ploughman sitting at an angle (one wheel in the furrow, the other up on the unploughed land), constantly looking over his shoulder to make sure the plough ran evenly and at the right depth. The horsepower was 8q or more, but the pace at three to four m.p.h. not much greater than that of the old two-horse team. As the shares move along so the wet earth falls back from the mould-boards in a continuous shining ripple – but for the visible heaving of the tractor it might be water they are cutting instead of the solid soil And at the end of each set of furrows (each traverse), the four ploughs are lifted clear, rocking the great tractor with their weight. (The old three-furrow double plough until recently slowly rusting a few yards from my door had all the solid immobility of a granite monument and weighed nearly a ton.) As the tractor turns on the 'headland' and backs ready for the new traverse, suddenly the four ploughs which were in the soil are reversed to the top, mud dropping from them. The four from above, their mould-boards gleaming unexpectedly silver, drop into the work position and dig down into the soil as the tractor moves forward.

All this work is to let the frost, the alternate wet and dry of winter work on the soil, so that in the spring the farmer may produce a good tilth for his seed bed. A heavy soil, ploughed early in winter, may look like strips of plasticine, but by the spring the clods will have broken and weathered, then, when the soil has dried, light harrows and rollers will achieve a tilth which earlier seemed impossible. But if the farmer goes at it before the land is dry, he will only puddle the soil into a crust which the seedling will never break through. If winter frost has failed to mellow the soil, he may have to 'force a tilth', but this means repeated expensive cultivations and is never more than a necessary make-shift.

Again, if the soil gets too dry before it is worked, it will form hard clods, which can then only be broken up by repeated working over with tined harrows, disc and zig-zag harrows and ultimately by using a roller. All of this emphasises what the textbooks say about the importance of 'timeliness', of carrying out – and completing – all cultivations when conditions are right. This in turn emphasises the advantages of the big farmer, cultivating 500 acres or more, over the small farmer, working one hundred acres or so with his son or with one other man. For the big farmer can concentrate men and machines, as a general concentrates his force at the point of attack.

4 *A spring-tined harrow drawn by a big four-wheel drive tractor and working direct onto the over-wintered plough furrows*

I remember, during a dry spell unexpectedly early in the spring, a smallish field on which four tractors were working simultaneously – all, incidently, wearing cage wheels, which are metal frames of the diameter and width of the tractor's great rear wheels, bolted on outside and thus nearly doubling the area of soil over which the tractor's weight is spread. They seemed too to be moving unusually fast, perhaps six m.p.h., and because the field curved in a peculiar way, there was an unusual scurrying and swirling, instead of the customary steady up and down, to and fro. The over-wintered furrows were broken down by disc harrows and then harrowed with spring tines (whose vibration at 'high' speeds helps to crumble the soil). The seed was then drilled in, the seed and its granulated fertiliser being placed by the same drill, the fertiliser a couple of inches to one side and rather more than an inch below the seed. The grooves were then neatly closed over. After that the light harrow went over the field again.

These general purpose seed drills may be twelve to twenty or more feet wide, with spouts six to eight inches apart and the seed must go in accurately to a depth of about $1\frac{1}{2}$ inches. For if the seedling germinates too near the surface it will die when the top layer of soil dries out, and if the seed goes down four or five inches it will fail to germinate at all. These wide drills have a marker running alongside and the tractor driver guides himself by the groove it leaves as he carefully makes his new traverse, so that the distance between rows remains constant. On the public highway these impressive creatures have to be transported end on, riding on a flat trailer behind a towing tractor.

After potatoes it may be unnecessary to plough at all before sowing corn (which of course means wheat, oats or barley), because potatoes may leave so good and clean a tilth that it needs only one working with a harrow. Potatoes themselves call for deep ploughing or cross-ploughing in the winter, down to nine inches or more and very often rotating hoes are used, power-driven by the tractor as it moves along. Potatoes inevitably still demand much hard work, although under the constant pressure to cut costs and cut corners, it is diminishing. First the carefully selected seed potatoes, generally from

5 *Combined seed and fertiliser drill. The disc sticking up against the sky on the left will be turned down on the return journey to cut a groove in the soil and on this the driver will centre his tractor when next moving from right to left. The tractor is wearing cage wheels, which spread its weight*

Scotland, are packed in trays and exposed to light day and night, until they have already sprouted. Then these 'chitted' seed potatoes must be carted to the field and the planting is a laborious business. A slow moving tractor may tow an attachment with, say, three bins of seed potatoes, behind which are three bare metal seats for men who sit all day, crouched forward as they feed the seed down shoots into the prepared furrows. These furrows are then earthed up (by splitting the ridges either side of them) and there is further hoeing before the rows are earthed up once more and eventually the tops of the plants meet and cover the whole field.

Completely automatic potato planters do exist, with one man slowly driving a tractor behind which the potato planter, with its bins of seed potatoes, automatically feeds them one by one properly spaced into the groove which has been dug for them, but it demands a pliant, loamy soil and evenly graded seed. Weeds too may be controlled by chemicals and chemical sprays are also used to combat potato blight – that disease which took its place in history by causing the great Irish potato famine of 1846 and hence contributed to the high Irish population of the United States.

When it comes to potato harvesting a machine with rotating spikes driven by the tractor's 'power-take-off' (the p.t.o. is the rotating shaft emerging just where the tractor's tail would be if it were an animal) will spin the spuds out of the soil (or there are elevator diggers which do better), but these potatoes still have to be laboriously picked up and boxed by men and women bending hour after hour with the wind playing on their backs just where it causes the most aches and pains. There are completely automatic potato harvesters – giant machines which are a joy to the lover of machinery – and they can be used by those who can afford them, but chiefly on light soils or the great flat expanses of black peat in the Fens, because they do have difficulty in distinguishing potatoes from stones or clods of earth, (except for the most complete types with electronic scanning devices). The reward for all this labour lies in the massive lorries driving away with their tons of (often) clean, yellow, uniformly-sized potatoes for the supermarkets.

Spraying may be done by light aircraft zooming perilously close to the houses as they make a trial run across the field. Then, as it skims above the crop, a cloud of fine spray bursts from the plane and is drawn like an unfolding veil across the field, before the pilot banks almost vertically to begin another run. This is flying as the pioneers knew it, but for the passer-by, who is nearer than he realised, it can be disconcerting.

Most crop spraying however, is still done by the tractor drawing a great tank containing anything between 50 and 250 gallons. Here again the tractor's p.t.o. will pump the fluid in a fine spray through holes in two great arms suspended by booms and stretching out (on the largest) up to thirty feet either side. These arms, modestly folded, can be reduced to quite reasonable dimensions when the tractor takes to the road and sprays are used not only against weeds and diseases but to kill off the leaves of peas and sometimes potato haulms before harvesting. When mild weather has encouraged excess growth of foliage in winter wheat, it also may be sprayed with sulphuric acid in the spring – or you may, to your surprise, find cattle in the cornfield grazing this 'winter proud' growth – to reduce the danger of too tall a stand of corn, which might be flattened by the slightest summer downpour.

The tractor, it will be clear, holds the centre of the stage throughout the whole year's unfolding drama, until it yields its place to the magnificent combine harvester as the arable year reaches its climax in the grain harvest. The land-worker begins his early day by starting his tractor engine and it runs almost continuously until many hours later he switches off and climbs stiffly from its high seat. The tractor itself starts its life with the hard work of ploughing, lifting and mucking out manure and in old age is relegated to the lighter work of hauling trailers on the roads. Its constant chatter, rising at times of great effort to a deep staccato bark is now the typical sound of the British arable countryside, eclipsed only at harvest by the majestic and unmistakable roar of the combine.

Tractors now come larger: 25 to 160 h.p., but 45 to 70 h.p. the most common. It is not only that they haul every instrument except the beet and perhaps potato harvesters and the combine, which are self-propelled; not only that their rear power-take-off with its range of regulated speeds will give independent movement to the towed instruments attached to it; it also has all the subtle strength of modern hydraulics. The tractor will scoop out all the accumulated farmyard manure for spreading on the fields; it will (towing or carrying the appropriate instruments) mow grass, cut hedges, turn or rake hay, bale straw and lift the great bales high on to a trailer. When it lifts great weights in front, it will have a concrete block slung for safety behind its rear wheels, and when it is straining a deep sub-soiling cultivator through heavy soil it will carry a weight in front to keep the front wheels from lifting.

The tractor will also distribute the nitrogen fertiliser which stimulates the growth of corn in the spring. Maybe an older tractor will haul the plastic sacks which are emptied into a hopper mounted behind and linked to the p.t.o. of the second tractor and this carries the hopper across the field, spinning out the fertiliser in a fine cloud of dust as it goes. In the spring too it may be found that frost has lifted the surface in some fields and loosened the roots of winter sown corn. On other fields heavy rain may have 'panned' the soil into a hard unfriendly crust. The tractor will be out with either roller or harrow, but when all the ploughing, preparation of seed bed and sowing is done, the high summer, apart from spraying, is unexpectedly the quietest time of the arable year, except for the wettest days of winter. Thus the winter barley, oats and beans will be sown between late September and late November and the winter wheat by January. The spring wheat, barley, oats and beans are sown between January and April; peas, sugar beet, cabbages and carrots between February and May – but all these dates depend on weather and soil, latitude and altitude, and above all on the weather being true to season; and one wonders whether at present the seasons are shifting.

Even before Christmas the fresh green of the winter sown corn will pierce the soil in straight lines of little spears, bringing the first promise of the new year's crops, for winter sown corn does not stay long under the earth. They will be followed in the summer by the broader leaves of the peas and brassicas and by the deeper green of the potatoes. The fields which were brown become green with growing crops and in midsummer the farmer watches his crops ripen, waiting for the weight of the ripening grain to bend the ears on their stalks, ready for harvest. 'The earth bringeth forth fruit of herself; first the blade, then the ear, after that the full corn in the ear' (*Mark*, iv, 28). Over years of experience the farmer has learned to accept with resignation the forebodings of his barometer, the freak storm which lays the standing corn or causes the oats to drop their seed (which they do all too easily). He must be resigned to persistent summer rain which will keep his machines from the ripe crop while the wheat sprouts in the ear, or to the excessively dry spring from which comes a carefully sown field of sugar beet in which only half the seed has germinated. Perhaps it is not unnatural that he accepts with much less resignation ill-conceived intervention of government or fluctuations in the cost of fertilisers, fuel and feeds, which he can no more control than he can control the selling price of beef and pork.

The country dweller knows then that evenly harrowed

fields in winter are already sown, while those which are deeply furrowed and highly ridged are to weather through the winter months before the seed bed is prepared for sowing in the spring. He knows too that June may be the quietest month before the hectic harvest months of July to September, starting with the winter oats and barley and peas in a good July, followed by winter and spring sown wheat, spring barley, oats and winter beans in August and September, spring sown beans in September and October, and then, on to the end of the year, the harvesting of sugar beet, cabbage and carrots. This is the arable farmer's year, but again all these dates vary with season and circumstance. What is certain however, is that to the corn-growing farmer and his men, indeed to the whole village, 'harvest' means the harvesting of the year's grain.

Then it is that the combine harvesters, like great mastodons, edge from their sheds and on to the roads, preceded, as the law requires, by tractor or Land-Rover with warning headlights blazing. The combine – a most expensive machine used only for a very short period of the year – is essential to any cereal grower and reduces by half the time taken by the reaper-binder, which was in general use until after the war. It can cut a swathe up to eighteen feet wide and deliver up to sixteen tons of grain in an hour. It can tackle crops which have been quite badly 'laid' by rain and wind and can leave stubbles of varying lengths, according to threshing conditions and the farmer's need of straw. The thing is a self-propelling factory, from the great 'reel' revolving in front, which passes the corn to the cutters (working like a barber's clippers on a somewhat larger scale), through the thresher which separates the grain from its husk (the age-old operation of the winter months), to the 'straw walkers' which deposit the decapitated straw and chaff in windrows on the ground behind as the machine moves forward.

6 *Combine harvesters at work. The wheat ears have turned down under the weight of the ripe grain*

The machine is strangely like a dinosaur, not only by reason of its size, but in the giant neck and apparent head which stretch out to empty the threshed grain from its storage tank into the succession of tipper-trailers drawn up alongside by the attendant tractors as the operation proceeds. A klaxon cries out too in apparent distress when it picks up a stone or hard clod, so that the driver may quickly stop the machine before awful damage is done to its complex mechanism. Some combines even have electronic sensors to tell the driver if too much grain is being lost and adjustment is needed. As the golden grain fills each trailer it hurries off to the drier, for if the grain is no longer dried by wind and sun in stacked sheaves it must be dried by artificial heat in the drying installations every cereal grower must have. Most grain is stored for sale when prices are at their best and it is an unhappy farmer who sells direct from the combine. Here again modern farming demands precision. If grain is stored containing twenty per cent moisture it will heat and ultimately spontaneously combust. At fifteen per cent it can be kept safely in bins provided it is occasionally turned (blown by fans through ducts from one bin to another, or aerated in metal silos by air blown through from the bottom). At fourteen per cent it is safe for long-term storage.

The scene on the harvest field has its beauty, but it would be wrong to see it in terms of the jolly pictures, left us by eighteenth- and nineteenth-century artists, of children picnicking amongst the sheaves which buxom lasses are stacking in the wake of the bewhiskered men with the scythes. Three great combines moving in echelon across a thirty acre field present a scene of majestic beauty, but their deep-throated roar obliterates all other sound and over the field, especially in dry weather, is quite a pall of dust. The men controlling these complex machines (it is the rear wheels which steer) are wearing masks (or should be); not just a gauze over nose and mouth, but the real thing with snout and goggles. The latest combines have enclosed cabs with air filters, but the air is not cooled, so that there is no great choice between the discomfort of a mask and the sun beating down on you hour after hour in an enclosed cab.

This is the period when men expect to work all the hours they can see, while the weather holds, all the days of the week, and

I *The ripening corn*

make little of sitting up all night with the dryer if the grain comes in wet, nor does the village complain as the murmur of the dryer goes on through the night. The grain is lifted from the trailers into the dryer, or the already dried grain intended for cattle feed is lifted into the granary, by blowers or by augers. Where once men heaved the sacks on their sweating backs, the end of the auger is now stuck into a grain container and the auger is a simple metal tube within which a sort of corkscrew rotates, carrying the grain from top to bottom of the tube. I have seen a child playing in the new grain as it might play in the sand, for even in this industrial age the whole village still participates in the excitement of the harvest.

The straw which the combines leave behind them may be burned by the farmer who keeps no livestock and can get no worthwhile price for it. This burning contributes some potash to the field's nourishment, but it is done simply to save labour and is a tricky business – and a dramatic one, for the darkening sky is lit by red and angry reflections, as it must have been by the camp fires of bivouacking armies in Napoleonic times. The field is then left scored in wide black stripes and the countryside is dramatic rather than beautiful, until the aftermath grows green or the harrow and the plough again turn the soil.

But if the straw is to be used, the combine will be quickly followed by the baler, a peculiar machine like a giant spider with arms wildly gesticulating through the dust. It collects the straw in the windrows and, to the accompaniment of a strange, irregular gulping sound, ties it up into solid bales – large, regular cubes whose compactness must be carefully controlled as the compression exercised by the baler will increase as the dew dampens the straw. A small sledge is towed behind the baler on which the finished bales collect and can be released in groups. This makes it easier to stack six or eight bales together and they are then often left for a time, both to dry and to release men and tractors for the more urgent work of getting in the grain from other fields. In the evening the bare field with its regularly spaced monoliths of straw looks like some straight-lined Stonehenge, but soon the stacks are lifted by the tractor's hydraulic loader and carried on big flat trailers to the Dutch barn. Here, amidst the smell of hot, dry straw, they are tossed on to an elevator and a steady steam of cubes moves diagonally upwards to the sweating man who stacks them as they land at his feet high up in the well-filled barn.

When it is all over, you can climb the stairs of the granary (noting its steel beams) and open the door on to the oats or barley stored for the cattle's winter feed – dunes of grain from wall to wall and from floor almost to the ceiling. And if your visit is in the early morning you may still see in the grain the footprints of the granary cat as it climbed up and over the great pile on its nocturnal duties.

2
THE PLOUGHED LAND

The mole works for its living by digging a tunnel through the soil, which it then patrols, gratefully eating such grubs or worms as come up into its tunnel or inadvertently drop in through the roof. Mole-draining similarly consists of driving channels three or four inches in diameter about two feet below the surface. This is done by bullet-shaped lumps of metal fixed on stout, knife-edged vertical beams or coulters, the whole apparatus drawn slowly forward by a massive caterpillar tractor. These 'crawler' tractors make poor progress on roads, so that they are most useful to the farmer whose fields lie comfortably around the homestead. They are slow and look as if their weight would compact the soil more than the heaviest tractor, but the reverse is true, because the weight is distributed along the whole length of the two caterpillars. Moreover, on a wet soil, you will see even a four-wheel-drive tractor with its wheels spinning, smearing the soil into a useless paste. This cannot happen with a caterpillar, whose ponderous strength – if the farmer has kept a machine of this sort – will without skidding pull out tractor, cultivator and all.

These mole drains naturally fall in after a few years and you will not drive far over the countryside without seeing stacks of tile drains waiting to be set in the fields. This is a much more elaborate and more lasting operation, undertaken by a specialist contractor with a machine which makes the most of a really big caterpillar tractor and of modern hydraulic controls. Usually a great vertically-set wheel armed with scoops cuts out a deep and narrow trench into which foot-long clay pipes are laid end to end. The lengths of pipe need not be joined but are surrounded for better drainage by a bed and covering of pebbles. The skill here lies not only in the whole lay-out of the drains, so that they empty into a ditch at the bottom of a field, but also in maintaining the correct gradient.

None of this is new, except the machinery and the occasional use of plastic or concrete drains. Some 150 years of experiment and experience have shown that if the land is to remain fertile it must be well drained. Quite apart from all the major arterial drainage work undertaken by public authorities, individual farmers are reckoned to have spent some £100 million (at the rates then in operation) on field drainage in the thirty years from 1939 to 1969. A well-drained soil makes more moisture available to the plant's roots than a saturated one, which will drown the plant; it encourages the plant to root deeply and – with deep cultivation – it improves soil structure and aeration. 'Wet land is late land', and most crops fetch better prices if they are ahead of the glut. A well-drained soil can be worked at a time when it would be folly to touch a waterlogged field, but more than that, dry soil heats up ten times quicker than water and it then takes much more heat to evaporate the surplus moisture as vapour, and that is why crops sown into a wet, cold soil are late in starting. A century ago Josiah Parkes discovered that by draining boggy land he raised the June temperature by 13°F at a depth of seven inches. The earliest crops come from

7 *The fully automatic potato harvester mentioned in Chapter 1*

well-drained sandy soils lying with a gentle slope towards the south or south west, or warmed in winter by sea breezes.

The maintenance of field drains and ditches is a winter job and the farmer must know where every drain enters the ditch. They must be inspected and if a persistent muddy patch or puddle in a field reveals a blocked or broken drain it must be opened up and dealt with. The quite different characteristics of different soils are revealed by their drains, because heavy clay and peaty soils hold water much more than light and sandy soils, which may parch their crop in a hot and rainless summer. You will find that the drains lying in light land begin to flow into their ditches earlier than those in the heavy fields; they flow gently and continuously, the volume of water increasing gradually as autumn and winter advance. The drains in heavy land, by contrast, will suddenly gush after rainfall, as the soil passes saturation point, will flow for a time at a great rate and then quickly fall away again to a trickle.

Heavy clay soils are hard to work, do not readily crumble into a good seed bed and will crack in dry weather, but they will not let the crop starve for moisture, because the fine particles in a cubic foot of pure clay possess a moisture-retaining surface area about sixteen times that of a cubic foot of sand. Liming improves heavy soils and they favour crops with strong roots, like wheat and beans, but not potatoes, whose roots lack push. On sandy and gravel soils ploughshares get worn down quite quickly, whereas on a soft, peaty soil one share may last a whole season. The best sandy soils will grow oats, barley, peas and potatoes, but the ideal soils, which suit nearly all crops – both in their fertility and supply of moisture – are the loamy oils, containing the right mixture of clay and sand. Chalky soils will drain freely and can be easily worked, but this means they are hungry for humus and manure and may set like cement if they are worked when they are wet.

The peaty and fen soils are of course full of organic matter and will grow most crops if they are well-drained, as our fens are today. They are particularly suitable for root crops, of which they produce vast quantities, but, by and large, the fertility of a soil depends on its depth and no amount of manuring can make a thin soil other than a poor soil, unsuitable for such crops as wheat, potatoes and beans. The soil in this country varies so much that one part of a field may differ radically from the rest of the same field and this its owner has got to know. Most of our good arable land is in the east of the country, from Yorkshire to Hampshire, with wheat, barley, potatoes and sugar beet especially concentrated in East Anglia and the Fenland, while the Midlands are heavy with beef and the wetter uplands of the west and north offer more grazing, especially to sheep. The old terminology still holds, for heavy land is 'three horse land' and a lighter soil 'two horse land' from the horse team needed to draw the old single-furrow plough.

The varieties of soil (and perhaps the varieties of temperament amongst the farmers) seem to stand out particularly well in the different methods of preparing seed beds for sowing cereals in autumn. Like the majority of the farmworker's jobs, this has to be tackled head-on, with long daily shifts and week-end work, while weather and soil conditions are right. One field of fairly heavy soil, a gravelly clay, is first ploughed. Then a harrow is taken at right angles across the furrows to break them up into a tilth. But this is a peculiar cultivator, with strong straight tines in two rows: steel spikes about six inches long. As the instrument is drawn forward through the soil, the tractor's p.t.o. causes each couple of tines to twitch to and fro against each other in a reciprocating movement, breaking the clods into particles. This is sufficient preparation of the seed bed.

In the next parish, on a pocket of real clay in the valley, a tractor strains and jerks to break up the stubble with a chisel-plough or cultivator fitted with strong narrow tines, intended for deep working. When this laborious job is done, the field will be worked with a disc harrow and then cross-harrowed with a zig-zag or 'seeds harrow', with two rows of tines, the rear row marching between the tracks of the front row. This sort of harrow gives the final touches to the seed bed and will also be used behind the seed drill, already waiting by the gate.

But only two or three fields away across the road is a fifty-acre field of light soil on top of chalk, well drained on a slope, and here the farmer is using a heavier zig-zag harrow, with curved tines ending in the shape which has earned them the name 'duck-foot' and he is using it for his first cultivation of the stubble. Behind a massive four-wheel drive tractor the wide double row of tines is ripping through the soil at a high speed, the driver sweeping round the field without bothering to back up at the end of each row – and again the seed drill, with both fertiliser and seed, is already waiting.

Between these two is a third field, also chalk, whose stubble has been ploughed, left to weather a little, then worked by disc harrows, after which a great flexible harrow – some twenty feet wide – with vertical teeth has proved sufficient. Down the hill on one side and up again on the other, the tractor moves at twice the speed of the one on the adjacent clay field and one cross-cultivation with a spring-tined harrow rippling easily through the wet soil is then enough to prepare the tilth for the winter seed.

There are also harrows consisting of thick steel tubes mounted on a frame held down by weights; and the disc harrows are particularly useful for preparing a seed bed after grass has been ploughed up. Those double rows of heavy metal plates, up to two feet across, running on their sharp edges are set at an angle to the tractor's line of march and the greater the angle, the greater their penetration. The basic difference between harrows lies in the depth to which the soil is to be stirred (and the deeper working instruments are strictly 'cultivators'), but their variety reflects the variety of soils and the age-old need to break up its surface sufficiently for the seed. And ploughs vary too. The lea plough turns long, unbroken furrow slices, its long mould-board gently twisting each slice over onto its side. The

8 *A massive, four-furrow reversible plough of the digger type*

digger plough's mould-board is much shorter, with an abrupt curve, throwing the furrow over in rough clods. Lea ploughs will only stir to a limited depth and digger or semi-digger ploughs are becoming more popular. Then there are sub-soil or really deep-digging ploughs, with tines which can be forced through the earth at a depth of two feet, breaking up the sub-soil without bringing it to the surface; and there are ridging ploughs with narrow shares and double mould-boards used for ridging up potatoes. With many variations in design, what all ploughs have in common is the beauty of a power which has evolved slowly over thousands of years.

After draining and working the fields, and the different sorts of soil, there is the question of manuring the soil. Gone now are the days when the prospective buyer of a farm carefully plumbed the depth of farmyard manure in the stables waiting to be spread on the autumn fields. Gone too the deep, rich, plum-coloured mixture which enthusiasts tell us could only result from a year's accumulation of horse droppings trodden into constantly added layers of straw, which was then loaded into one-horse tumbrils and spread over the fields – a job which might go on for so long that the steam from the final loads was visible in the chill December air. To the many still working on the land whose hearts linger with the departed horses, cow droppings are a poor second best. Yet when heifers or cows have over-wintered in covered yards the accumulated cow-pats and urine, soaked up and trodden into successive layers of straw, still produce manure which greatly improves the tilth of the soil and, like all humus, helps it to hold moisture. Moreover, its contribution to the soil's fertility lingers on to a far greater extent than that of the chemical fertilisers which arrive in their antiseptic plastic sacks.

The yards are 'mucked out' in the spring and it is important that so far as possible the manure should be protected from the winter rains, which would wash out the valuable nitrogen. Many farms of course now carry no livestock at all and, as the 'Strutt Report' of 1970 pointed out, some soils are now suffering from 'dangerously low levels' of organic matter. To the specialist 'indoor farmer', who keeps great numbers of cattle, pigs or poultry on acres of bare concrete or on slatted floors and who cultivates no arable land, this manure is at best waste and at worst a problem, something to be disposed of in liquid form as 'slurry', the cause of what one writer has called 'the age of effluents'. But on mixed farms you still hear the bark of the tractor as its scoop heaves the rich mass on to the trailer to be piled ready for next autumn's ploughing, or direct into the distributor, which will spread it over the fields. This is a peculiar metal trailer with a slatted floor which moves from front to back, shifting the load with it and at the back there may be two sets of cylinders which will shred the muck and then fling it out in a storm of dark flying lumps behind the advancing tractor. It may seem a strange topic to evoke enthusiasm, but there is something about the sight and the full-bodied smell of a heap of well-strawed, well-rotted manure which epitomises the richness of good land and of what it yields.

But there is no denying that muck (nowadays mainly used for potatoes and sugar beet), because it is spread in ton loads, means work, even with modern machines. And the food-producing industry runs on the narrowest of margins, unlike, for instance, cosmetics, where money counts so little that more may go into the mere wrapping than into the contents. This has worked in favour of the artificial fertilisers, which are cleaner and less laborious to handle, but the growing concern about soil structure and the increasing cost of nitrates and of imported potash may soon balance the extra man-hours involved in

9 *Loading farmyard manure into a muck spreader*

spreading farmyard manure. Nitrogen, as every gardener knows, promotes the growth of the plant's stem and foliage; phosphates develop a good root system which enables the plant to utilise nitrogen without weak, lanky growth; potash helps the leaves in their job of producing starch and sugars and hence the grain, fruit or tubers for which the plant is grown. The great joy about artificial fertilisers is that they come with these three basic constituents mixed in exactly the right proportions for the required crop; for spring or autumn use; or for the state of the soil and the fertiliser residues left over from the previous crop. Thus potatoes, sugar beet and the market gardener's crops will welcome a high proportion of potash with nitrogen and phosphate; spring sown cereals need a preponderance of nitrogen with phosphate and potash; phosphate and potash only go into the seed bed for autumn sown cereals and easily soluble nitrogen is then given as a top dressing in spring and strongly improves the crop's early growth.

Artificial fertilisers can also be so precisely controlled that for instance, a deficiency of magnesium can be made good by spreading a few pounds to the acre. Spectacular results follow when very small quantities of fertiliser slide down the same spout with the seed; equally, artificial fertiliser can be drilled together with peas, but just to one side of the row, since peas will not tolerate actual contact. There is no guessing or approximation about this: optimal rates for the use of fertilisers have been worked out, taking account of the cost in relation to yield, and mixtures can be obtained which reflect the fact that wheat needs more phosphates in wetter areas. In south-east England fourteen units may suffice to the acre, where the north and west require twenty-four and Wales and Scotland as much as thirty-four (one unit is 1.12 lb or 1 per cent of one hundredweight in pre-metric terms).

Finally, different crops, like garden plants, tolerate different degrees of acidity. On the pH scale, 7 is neutral, readings above 7 are alkaline and readings below are acid. Barley is the most sensitive of the cereals to acidity. It prefers 6.5 pH, will give reduced yields at 6 and will fail at 5.3. A patchy crop may be due to patches of acid soil and what is too acid must be limed. Lime too will improve grass for grazing and it can be spread by the same distributor which flings out the lumps of farmyard manure.

With his soil in good heart, the farmer turns his attention to the pests, diseases and weeds which may attack his crop and rob him of his harvest. To read the list of pests, funguses and viruses which attack crops is as depressing as to see in a *Home Doctor* a reminder of all the ills to which the flesh is heir (and livestock are equally prone to disease, particularly when great numbers are concentrated in a small area by the 'factory farming' on which we now depend, but which would have caused our forebears' eyes to start from their sockets). The weapons to the farmer's hand are immune varieties; seeds immunised against certain diseases before planting; chemicals applied as sprays or powders to check air-borne and soil-borne diseases and pests, as well as weeds; clean husbandry and crop rotations.

'Take-all' is a fungus which attacks the roots and eyespot a fungus which attacks the stems of cereals. The eelworms ruin oats, potatoes, beet and peas and are an even greater menace to agriculture than insects. Fortunately eelworms do not at all readily move from one field to another. The defence against all three of these menaces is therefore to break the continuity of the crop on which the menace flourishes and this is vital with the cyst eelworms, because once a cyst population gets up to danger level, the field is likely to remain almost permanently infected. With potato cyst eelworm the population builds up so quickly if potatoes are repeatedly grown on the same field, and the damage to the crop is so great, that it is better to err on the side of safety and to use a field for potatoes only once in every four or five years, although the *early* potatoes manage to complete their growth before most of the eelworms have grown up. A change of crop for one year is otherwise generally sufficient (except for eyespot which needs a break of several years), provided the field is cleaned so that no couch-grass, Yorkshire Fog or self-sown ('volunteer') wheat or barley is left to act as host. The farmer who nowadays specialises in growing only cereals year after year is wise to plough in or otherwise clean and cultivate his stubbles as soon as the harvest is gathered, and to sow his new crop as early as he can. Repeated heavy 'discing' both works the chopped straw into the soil and exhausts the couch-grass rhizomes. A new development which seems to check 'Take-all' is to kill off the stubbles and their aftergrowth with the lethal chemical Paraquat and then to drill the new seed straight into the unploughed soil, using a drill which makes a narrow slit for the seed.

In all this, let me take one example of how the man on the tractor applies the results of detailed research. If successive cereals are grown, 'Take-all' usually increases to a peak and then declines. Barley will endure the rising and the peak years better than wheat, but when the peak danger is passed a more remunerative crop of winter wheat can be tried, using a wheat variety resistant to eyespot, so as to break that fungus' cycle as well. If the farmer changes to another, non-cereal crop, 'Take-all' is immediately reduced, but then quickly builds up to a higher peak than before the change! Say that before the break the cereal yield was 25 cwt per acre; the first crop after the non-cereal break may be about 40 cwt, but the second and third only 20 cwt each, before the yield settles down to the original 25 cwt. The average over four years is only about 26 cwt and one year's cereal crop was sacrificed to make the break, so that the whole exercise is only worth it if the non-cereal crop was a very remunerative one.

Air-borne diseases are yellow and black rust in wheat and mildew in wheat, barley and oats – and as they are airborne you are at the mercy of your neighbours. All you can do is to be meticulous in destroying weeds or seedlings sprouting from spilled grain, as they would harbour the spores over the winter into the next cropping season. The plant breeder helps here with resistant varieties, but there is no final victory in this battle, because the fungus too mutates and produces new strains. Spraying goes on in spring and summer to control leaf diseases like potato blight and to kill weeds, which would compete with the crop, interfere with harvesting and harbour pests and diseases. It is said that about one ton of insects could be collected from every twenty acres of ploughed land (or from as little as five acres of grassland), but the plovers we so often see on the fields eat wireworms, leatherjackets, chafers, slugs and snails, and rooks, jackdaws, magpies, starlings, seagulls and other insect-eating birds all eat at the same table.

Despite man's efforts and the appetite of birds, crops do still get lost and no sight is sadder than a field of upstanding wheat all black with rust where there should be the gold of the harvest. You cannot help thinking of the amount of care which has been taken and is now wasted and although things can go wrong in factories through no fault of the manager, it is still easier to control the environment in which your product comes into being indoors than it is outdoors. And this comes home as forcibly with sugar beet as with any crop, because of its value if it is successful, because it demands a field like a billiard table, and because even then the farmer is paid according to the sugar content, which is virtually beyond his control.

Consider the sugar beet, which we are growing increasingly since its first introduction in 1920, because tropical sugar cane no longer meets the world's needs, although this is a fairly localised crop, needing a deep soil. The field is cleaned in the autumn, dunged with about ten tons of farmyard manure to the acre, well and deeply ploughed before the severe weather starts. There may be a 'plough-pan' or hard, impervious layer in the soil just at the depth of the usual plough furrow and this must be broken up by sub-soiling – using a tractor of sufficient strength to draw tines or a digger plough through the soil below the usual depth. Sugar beet requires sufficient lime and in the spring the frost-mellowed plough furrow is worked into a fine and firm seed bed. First a tined harrow goes across the furrows working to a depth of four or five inches, then lighter harrows take over and ultimately the soil is rolled with a Cambridge roller (which unlike the garden roller has a corrugated surface and leaves the soil in little ridges, resistant to the 'panning' effect of heavy rain).

The seed is drilled as soon after the middle or end of March as the soil permits and it must go no deeper than three-quarters of an inch into a moist tilth. Sugar beet is peculiar. It produces a fruit containing several seeds, which germinate into a group of new plants and in the past an immense amount of laborious but delicate work was put in by gangs of men working down the rows to reduce the clumps to the ideal of about 33,000 single plants to the acre. 'Singling sugar beet' is still a phrase which rings, particularly in the villages of East Anglia and Lincolnshire, and you will still see the work done by hand, but nowadays fruits are prepared containing only a single seed and these are made up like pills or pellets, so that precision sugar beet drills can easily deliver them at regularly spaced intervals. A flat roller wheel runs behind each seed tube and the drill also neatly places a weed-killer on the soil behind the roller. A light harrow may then go over the whole field, leaving it so neat that it will not fail to catch your eye.

If further thinning is needed the thin-wheeled row-crop tractor, treading carefully between the rows, will draw a machine with hoe-heads on vertically mounted drums which rotate, knocking off plants in the rows at intervals as they turn. The number of hoe blades can be varied according to the density of the seedlings and this is worked out by using a frame which enables one to count the seedlings in a given area. In 1946 only one per cent of sugar beet was not harvested by hand, laboriously and painfully – for the tops had to be chopped off with a sharp knife or 'hook' wielded by cold fingers. Now the machine has completely taken the place of the vanished labourer and self-propelled harvesters spin one row of roots out of the ground while topping the next row in readiness to be jerked out, picked up, cleaned of loose soil and lifted into a tank which can be emptied into a trailer drawn alongside. Or the job can be shared by two less massive machines. Some sixty-eight man- or woman-hours per acre went into hand pulling, knocking and topping. The complete harvester reduces this to twenty-six, but it does leave an appreciable proportion of the precious crop in the ground. Surprisingly, for a complex and heavy machine, lifting as unromantic a vegetable as the sugar beet, it gives in operation the impression of a moving fountain. The beets go flying up to curve forwards in a continuous stream into the tank, while the green tops are also flung into the air, before falling in a graceful plume backwards on to the ground.

How precious is the crop? Sugar beet if left to itself grows its leaves one year and sets its seed the next. If it is checked by bad weather in the spring, it may fail altogether to germinate properly or it may 'bolt' and in its first year produce unwanted seed at the expense of the desired sugar in the beet. It may suffer from 'speckled yellows' due to manganese deficiency and this calls for ten pounds of manganese sulphate per acre, sprayed on to the leaves as soon as they reveal what is wrong. But the whole leaf may turn yellow and this means that it is suffering from 'virus yellows', spread by aphids or greenfly. British Sugar Corporation fieldsmen are out to check for concentrations of these insects and spray-warning cards are sent to beet growers, but the crop which is heavily infected in July may lose half its sugar. All this sugar beet is grown under contract with the factory which is to refine it, and to check sugar beet eelworm – very prevalent on the Continent – the contract will specify a two-year break before another sugar beet or brassica crop may appear on the same field and all infected fields are scheduled. The factory buys at a fixed price per ton of washed roots yielding sixteen per cent of sugar – more for every one per cent above and less for every one per cent below sixteen. Here, as an example, we have followed in intensive detail the intensive work put into what should be a good cash crop but which, in the last resort, escapes the farmer's control. At least, like potatoes, sugar beet leaves the field well manured, cleaned and deeply worked for other crops, and its pulp and leaves can be used to feed livestock.

The British Sugar Corporation's factories are most easily found in the eastern parts of the country where the bulk of the beet is grown. The massive cylindrical silos in Norfolk near the Wash are visible across miles of flat, black, hedgeless Fenland, rearing up into the great arc of the Fenland sky and only dwarfed – but magnificently dwarfed – in memory by the venerable pile of Ely Cathedral towards the south of the Fens. This factory in the unpopulated fields is like a city in an otherwise empty plain. Beside the storage silos are the buildings housing the great cylinders in which the beet is rendered down and the sugar extracted; there are lorry parks, unloading bays, elevators, overhead conveyors and chutes, all laid out for the mechanical handling of the heavy and bulky roots, and of their pulp residues, in the minimum of time. The ponderous lorries arrive, each by appointment, in a continuous stream during the period of maximum pressure as the beet is lifted in the autumn and early winter. And then, strangely, this city in the unpopulated fields is deserted again: the factory closes when the last of the year's harvest has been processed.

The soil has been drained, manured and cultivated, freed, so far as possible, of both annual and persistent weeds, pests and diseases. There is one basic rule of good husbandry, to leave the land better than you found it. 'In the soil lies all that remains of the work of countless generations of the dead.' (George Henderson, *The Farming Ladder*.)

3

GRASSLAND AND THE GRAZING HERDS

Until recently, the fashion was to make the cow walk while the man stood still. But now the cow's amble is too slow for modern needs and the latest idea is to place her on a turntable, so that she can be moved more expeditiously on-stage and off-stage again after milking. Far behind us are the days of the cowshed, where Buttercup and Daisy stood placidly in their stalls while human fingers squirted milk from their teats into buckets. The modern milking parlour has something of the atmosphere of a church, in the calm and decorum of what might be termed its ritual, for a thirteen hundredweight cow which is giving five gallons of milk a day must not be hurried or worried and a disturbed cow will not 'let down' her milk. The men stand in a pit, so that the cow's teats are at eye-level; the cows file quietly into their milking stalls either side of the pit (or the turntable places them there); a small ration of palatable cow-cake (in effect a lollipop) is dropped from an overhead container into a receptacle in front of each nose; hanging above each man is a rubber brush which he pulls down and through which warm water containing a mild disinfectant flows on to the teats as he washes them; then he applies an electric milking cluster to each cow, and eight or twelve or more cows will yield their milk simultaneously to the pulsing of the milking machine.

The routine provides for the 'fore-milk' to be drawn at regular intervals by hand, so that the cowman may see any signs of blood or lumps, which would be a symptom of mastitis (or inflammation of the udder), but when the milking machine is working all the milk from all the cows frequently flows through its tubes straight out of the milking chamber into a great, cooled tank behind the wall and from there is pumped straight into the dairy's tanker-lorry. Buttercup and Daisy probably still have their traditional names and are personally known to the cowmen, but most likely each also wears a plastic neckband bearing her number, and this number appears on the blackboard in the tank room on which her yield is regularly chalked up. Her parentage too will be recorded back through very many generations, for milking herds are high-pedigree cattle, most commonly now black and white British Friesians. But she will not know her father, because artificial insemination is now almost the rule (and her father might indeed have died before her conception, for the semen of prize bulls is kept in deep freeze).

On the human side of the milking parlour there is a machine room and very adequate washing facilities, for hygiene is scrupulous, and probably a small office or rest room, for the cowmen spend forty per cent of their time in the milking chamber. Cows are generally milked twice daily, but some of the highest yielders are worth milking three times a day. On the cow side of the milking parlour are a commodious straw-covered area where the cows can lie down, but also bare concrete passages, easily hosed down, from which cow-droppings can be scooped up by tractor. All along the outer wall are the feeding troughs, which can be conveniently filled

10 *Part of a milking parlour. In this, each cow's contribution is visible before being pumped into the common cooling tank*

from outside, each cow being allotted sufficient space by the railings through which she has to push her head to feed. The whole of course is roofed, but not fully walled, so that, while the animals' own body heat maintains a satisfactory temperature inside, there is always plenty of fresh air. Isolation pens for any whose health is in doubt complete the cow side of the accommodation and calm is the key-note, even when the building contains 150 or more cows at a time. A milking cow will drink about fifteen gallons of water a day and a further fifteen gallons per cow may be needed for cleansing and cooling, so that a great deal of water is needed and may justify a large farm building its own reservoir.

There are of course many variations on this, but with all livestock there is an increasing tendency to design storage, feeding and bedded areas as one unit. On the whole, it is found that there is less labour if cows are not kept in stalls. The food may be placed in the feeding troughs quite automatically, by auger or hopper, or, instead of the feed being brought to the cow, she may go to her feed. Generally she only spends the winter indoors and from spring to late autumn in the fields, but there are cows who spend their milking lives indoors ('zero grazing', which means that all their food is brought to them from the fields).

Grass is the basic food and is one of the farm's most important crops. It is after all, mainly from grass that we get our milk and cheese, wool, beef and mutton. Grass also happens to be one of the most difficult plants to establish successfully, and that goes for the garden too. It needs more careful preparation of the soil than any other crop: a tilth and weather conditions which are perfectly adequate for wheat or roots will fail to germinate grass or clover, and the seeds themselves are so small, containing so little nourishment, that when they do germinate, the seedling plants die off all too readily in a dry spell. For generations, putting down a ploughed field into good permanent pasture was one of the most unpredictable operations on the farm. Only since about 1940, with the new machinery, modern soil analysis and precisely tested and appropriate seed mixtures has it become possible to lay down a grass sward with predictable results. The National Institute of Agricultural Botany in

Cambridge provides a precise specification of 'herbage seeds' and certifies their quality. (It actually publishes estimates of the digestibility or 'D value' of different grazing plants.) So it is now a routine to establish a grass sward, for grazing or cutting for a few years (a 'ley'); or as a permanent pasture. A short ley can take its useful place in the crop rotation and when ploughed up again leaves a good tilth.

There has been strong feeling about 'taking the plough round the farm', but now it is agreed that, provided they are equally well cared for (which was often not the case), permanent pastures are as good as newly sown swards and that there is no virtue in ploughing them up unless it suits your rotation. The permanent pastures or long leys with their firm base bear the weight of heavy stocking with beef 'stores' (that is, cattle being fattened for the butcher), while the shorter leys often carry the dairy herds. The finest permanent pastures are of course in the 'Shires' (the Leicester, Warwick, Northampton triangle). It is only the recent trend towards more and more cereals which has sent the plough through many of our permanent pastures, particularly in the eastern counties, except where the water table is high, as in the Essex Marshes.

Until fairly recently twenty-five or more species of grasses and clovers were mixed together when sowing for permanent swards or long leys in the hope that some at least would be happy; some would flourish in wet and some in dry seasons; some give the cattle an 'early bite' and some go on well into the autumn. Where before all was hope and guesswork, now there is precision. We know which varieties suit which conditions, how competition between plants works, and choice has been simplified by specially bred varieties. There are consequently, in the wetter west and north, pastures in which Perennial Rye-grass dominates and in the drier east and south those dominated by Meadow Fescue. On the lighter soils, which tend to dry out in a dry summer, you are likely to find Cocksfoot, but on the heavier and wetter soils, Timothy may take its place – and as Timothy is peculiarly palatable, care must be taken that it is not overgrazed. In addition, there is White Clover, which as a turf-former has no equal and also usually the late-flowering Red Clover. Both offer more protein than grass. In the drier parts deep rooted Lucerne (Alfalfa) may be the main plant.

For permanent pastures then the mixture of seeds may well be (although there are many variations): White Clover S100 or Kersey, late-flowering Red Clover, Timothy S48, Cocksfoot and either Perennial Ryegrass S24 with S23 or Meadow Fescue S215 with S53, all in appropriate proportions – and this, perhaps tedious, detail pays homage to the research it attempts to summarise, which has made this no longer a matter of hit or miss. In the shorter leys, under three years, there is greater variation, because their purpose varies. A one-year ley may be intended to give the soil a rest, or to provide short-term abundant pasturage, or a heavy crop of hay. But, whichever the purpose, the appropriate mixture of seeds is known and for a single year Italian Ryegrass (which only lasts two years) will be included, because it provides leaf for so long before it runs to stem and flower. These temporary leys are generally sown under a cereal 'nurse' crop: a spring cereal is drilled as early as possible and as soon as the cereal seeds are put in, the soil harrowed and rolled, the same is done for the grass seeds. The cereal is then harvested and the grass takes over from the stubble.

Having got your grass you have got to fertilise it like any other crop and the effect of different fertilisers on permanent grass has been studied continuously since 1856 at the famous Rothamsted experimental station. A dressing of nitrogen will advance growth in the spring and it is worth applying nitrogen between periods of grazing, while phosphates prevent the grasses swamping the clovers. If deficiencies in lime, phosphate and potash are made good and nitrogen is then given, the poorer sward 'communities' will become better balanced. There is a 'do or die' method of cruelly harrowing the poor grass of old pastures which can also be risked with cereals that are doing poorly, for they too are only cultivated grasses, with the same fibrous roots. If grass roots are deliberately damaged by the harrow, they are often stimulated into 'tillering', or growing more roots, from which the plant can only benefit. As an alternative to this physical approach, there is the purely chemical one of killing off old pasture by spraying it with a grass killer, particularly on steep hillsides which cannot well be ploughed. A rotary cultivator will then produced the tilth into which the seed may be sown for a new sward.

The trouble with grazing cattle and sheep is that they both tread and leave their droppings on what they eat. This is nature's way and cow-pats make manure, but the cow will avoid soiled grass and will also find no food if the soil is 'poached' by too many hooves passing too often over the narrow passage by the gate. Everything then turns on concentrating the stock on one part of the pasture at a time, so that what is available is fully grazed and the rest can grow unspoiled. And the numbers grazing must be regulated according to the growth of the grasses and clovers. This, together with care of the sward, is the pith of good management. On this – outside factors apart – will depend the income derived from the whole enterprise.

The stock will be glad to get on to the spring grass, but the early grass must be given its chance, because each new leaf helps to produce another leaf. Between mid-May and June or July herbage grows fast and the density of the stock is increased

11 *Friesian cows grazing up to an electric wire*

to keep the grass closely grazed, so as to prevent the more vigorous grasses crowding out the finer grasses and the clovers, and to postpone their natural tendency to set seed. It is at this time especially that the passer-by admires the deep lush green and the ability of cattle to wrap their tongues round it, to tear it up and to digest it with enjoyment. The grazing pastures may be divided into a series of paddocks, and the herd left two or three days on each. If a dairy herd has, say, twenty-one paddocks, it can spend a day on each and the grass will have three weeks to recover before it is grazed again. Or the movable electric wire will divide the pasture into strips. This gives a mild shock to anyone touching it, but you will see the line of noses grazing right up to the wire, grazing close, but not going beyond until the wire is moved; and the system responds very flexibly to the rate at which the grass is growing, which depends on the warmth and the rain.

So much for the main natural food from 'first bite' in spring to late autumn, but what about the winter? The traditional answer has always been hay, and this is still the most popular answer. Fields intended for hay are generally harrowed and rolled in the spring as soon as they are dry and are not grazed until the crop has been taken. Then the aftermath is grazed and incidentally if a field is mown year by year and only the aftermath is grazed, the sward will differ from one which is grazed systematically, because the taller hay grasses will shade and finally suppress the smaller grasses and the White Clover and you will be left with only the stronger species, like Perennial Ryegrass and Cocksfoot. Wise management therefore mows and grazes in alternate years, so as to maintain a better balance. The best *quality* of hay or silage (the alternative to hay) comes from cutting just as the plants are beginning to flower; the greatest quantity when they are in full flower. If the hay is cut too late it will have lost a lot of protein and (to quote the text-book) 'nothing done subsequently can retrieve a position already lost'. But there are two problems here, the problems with which the farmer wrestles all his life: can men and tractors be available at the right moment and, in our changeable climate, will there be a dry spell when he must have it dry, and will the dry spell continue long enough?

Hay lasts through the winter because it is dry and silage because the plants are killed as quickly as possible by smothering them, so that they get no oxygen. What the stock eat in silage is preserved by the acids produced by fermentation of the sugars in the sap. Any crop which is suitable feed for livestock (maize, kale, pea haulms) can be made into silage, provided it contains enough moisture and sugar to ferment properly. The art of making good silage lies in foreseeing and evading conditions liable to result in the wrong type of fermentation, and in avoiding excessive loss in the field and in the silo. With hay too the art is to get it safely into the barn, having got rid of 75 per cent of its natural moisture without losing too much of the nutritious leaf. (Hay containing 35 per cent moisture will rot, at 30 per cent it will mould and heat, at 25 per cent it can be baled, but has got to lose another five per cent before it can be safely stored in open-sided Dutch barns.)

When the mower has cut the grass for hay – and anyone who has not smelled freshly mown hay on a summer's day has a right to complain of ill-fortune – then the sooner it is turned, and the more often, the quicker it will dry. There are various hay tedders and turners, of which some are more gentle with the hay than others, but the finger-wheel swathe turner is one of the farm's most beautiful instruments. It is raised and lowered as usual by the tractor and its four wheels – like giant bicycle wheels with delicately turned spokes – are moved by contact with the ground when the tractor goes forward. As they rotate they gently pick up and turn the swathe. There is also a 'crimper', used once only, which passes the cut grass between two rollers and, by simply bruising the stems every few inches, greatly speeds the task of drying the hay before it is baled. Unfortunately the baler inevitably shatters the now dried leaves and on a hot day the loss may be considerably greater in the afternoon than in the morning, so that only experience is a safe guide as to the right time to bale and cart. The latest machines turn out big bales, each weighing half a ton, and less tighly packed than those of the ordinary baler. About two tons of hay comes off an acre of meadow.

12 *The finger-wheel swathe turner*

Grass for silage may be cut by the ordinary mower or by a flail harvester, which is essentially a rotating axle with a large number of knives mounted on swinging arms, decently covered by a hood. The chopped grass flies up a funnel and eddies like solid smoke into a trailer, towed behind or alongside, which collects it between high wire sides. The flail cutter delivers the grass long, lacerated or chopped. The last two ferment better, because the sap is spread in the process, but cattle need some long-fibred material in their feed (from this or another source) if they are not to become costive. Tower silos – you see them from afar on the horizon – are suited to chopped grass, which can be blown up into them, while walled clamps are most easily made with lacerated or long grass. Essentially these are piles, preferably between walls, rammed tight to expel the air and then covered with plastic or tarpaulin sheets to keep them air-tight and rain-tight – old tractor tyres, bales of straw, breeze blocks or anything the farmer has handy will hold down the covering. Inevitably there is loss through the process of fermentation, through putrefaction where the air has got in or where the tractor's wheels have muddied the grass, and in the liquid – smelling to high heaven – which drains away as the herbage consolidates. In a walled clamp the loss may total around twenty-five per cent.

With both hay and silage the method used is less important than doing well whatever is done and with both the difficulty is to know whether what one is feeding to one's stock during the winter is really nourishing or not. One very experienced farmer comments that one winter cattle fattened on silage and barley gained 3 lb daily in weight, but in the following year, on what seemed an identical ration, only gained 2 lb a day.

Beans and peas are more often grown for stock feeding than for human consumption, at any rate outside the vegetable growing areas, and cattle will eat a fair amount of oat straw – the rest, with wheat and barley straw, they use as bedding. In addition they need succulent foods, such as cabbage, kale, turnips, mangels or 'chat' potatoes (which are the small ones that would be rejected by the supermarkets). In the past, it was reckoned a great discovery that stock could be winter-fed on roots and Lord 'Turnip' Townshend is of legendary fame, but

II *Harvested cornfield. The stacked bales of straw make it look rather like Stonehenge*

then he did not have to pay his men much by way of wages, and roots mean work.

This grass eating brings us to the advantages which the ruminants (cattle, sheep and goats) have over us, who can ruminate in the mind but not with any satisfaction regurgitate from the stomach, and this leads us to the organs they have evolved which enable them to chew the cud. There are four stomach-like compartments, but three are more or less enclosed by the rumen and it is the rumen which (in strictly non-scientific terms) enables these animals to eat the same food twice. They store great quantities of vegetable matter in the rumen; and its capacity is enormous – up to fifty or sixty gallons in adult cattle. Here the fibrous cellulose (which we could not digest) is broken down by bacteria and protozoa and some parts of the food are digested without more ado, but the solid vegetable fibre is regurgitated into the mouth. The cow then settles down to chew the cud with her molars, grinding it down and mixing it with copious saliva. When she swallows it the second time, the cellulose has been well broken down, 'making possible a better attack by the normal gastric and intestinal juices'. This is the secret of how ruminants deal so well with fibrous foods. Horses and rabbits also cope reasonably well by an enlargement of part of the large intestine, but the complex ruminant stomach is the best converter of grass into flesh and milk.

An animal is not contented if its stomach is not full and if it is not contented, it will not thrive. But the 'stomach' takes quite a bit of filling in adult cattle and sheep, as we have seen, and as with us the amount eaten depends on how palatable the food is, except that dairy cows have a marked advantage over fattening 'stores' in being able to eat as much as their yield requires. Grass is highly digestible until early June and it is their natural feed, but only the farmer's experienced eye can assess what value his animals are getting from their industrious grazing. The natural rhythm is to graze and then lie down and chew the cud. You do not want cattle on their feet all the time searching for fool; if they are restless it is time to move them to fresh grass. It is one of this book's recurrent themes that a man's skill is being constantly exercised and that he is calling on the judgement born of experience more often and continuously than he is probably even aware.

Cattle like it succulent and juicy and among the concentrated artificial feeds presented to them they prefer flaky, crushed material in pellets and cubes to dry, dusty meals. During spring and early summer really good grass alone will give a cow her maintenance and enough over to yield up to four gallons of milk a day. But some pastures will give only enough for one, two or three daily gallons in addition to maintaining the animal's own health and then her rations will have to be supplemented by hay, beans, peas, crushed oats, sugar beet pulp, groundnut cake, or similar proprietary concentrated foods, and the same claim can be made by a cow even on top quality grass if she is giving more than about four gallons. On full winter rations she will need hay, oat-straw or silage to make her feel comfortably full, together with the concentrated foods. Kale is grown for cattle particularly in the south and west (Marrow Stem Kale before Christmas because it contains more digestable dry matter, but as winter advances Thousand-headed Kale because it withstands the frost better). Silage can be delivered automatically from a silage tower to the feeding trough, but if the cow feeds herself from the clamp, the face at which the herd eats must be large enough to allow each cow to do so quietly and in comfort. Then, in good appetite, she will eat something like 100 lb of silage a day, which means about seven daily tons of this feed, which is wetter than hay, for a herd of 150 cows. Feeding comes down to getting the right balance between bulky, low-energy foods and the concentrated high-energy, highly digestible but correspondingly expensive feeds which provide us with our milk or meat. The better the quality of the hay or silage or other bulky feed, the smaller is the need for the more expensive concentrates.

The old dairy herd of twenty to forty cows has now almost disappeared and the tendency seems to be towards herds of about 150 cows. Larger herds of, say, 250 cows are by no means unknown, but in many ways it is easier to manage two good-sized herds than to join them into one. One man, with the right equipment and well-planned buildings can milk up to 120 cows, but this means that he has got to keep his eyes wide open for any signs of illness (mastitis and sterility are the most common troubles) and he must also not fail to notice when the cow (or the heifer who has not yet joined the milking herd) is on heat and is ready for artificial insemination. (Here it is useful to have a bull around, as he will notice, before the cowman, when a cow is on heat.) Antibiotics have prolonged the lives of domestic animals as they have our own and in the last few years the working life of the dairy cow has doubled. Where it used to be only two lactations, of about ten months each, it is now four, and allows cows to reach their peak yield. In the past, virtually every heifer born had to be drafted into the milking herd simply to keep it up to strength, but now it is possible to select the most promising and those selected will live about six years. To buy in milking cows is unwise, not only because of the real risk of introducing disease, but also because the stress of joining the herd often proves too much for the animal. Cattle may seem bovine but they are not unaware of their neighbours and I remember the story of two similar herds which got mixed on a road and, as their drovers could not separate them, had to be left together overnight in a field. By morning they had separated themselves.

Spring calving is natural, but we expect milk all the year round and the manager of the dairy herd exercises considerable skill and judgement to maintain a steady flow, bearing in mind, in forming his equation, that cows yield half their total in the first three months of lactation; that milk yields automatically rise in May and early June when the grass is at its best; and that nature provides a much greater supply of milk for calves born in the autumn, although he also has to remember that the autumn calving cow cannot go out to grass, but must be fed on preserved foods. A dairy cow should be inseminated so that she is given a rest of six to eight weeks between ending one lactation and giving birth to the next calf, whom she will have carried for about 283 days.

This calf, when it is born, will be an animal bred for a single purpose. Until about 1950 a farmer produced a single breed, according to his choice and generally according to the part of the country he was in: the cows would give a useful amount of milk and the steers could be fattened to make saleable beef. The last twenty-five years have tilted the balance towards specialist dairy breeds, because of the good profit on milk, and 'dual purpose' breeds have almost gone out. Recently it has become clear that the Friesian cow will give the highest milk yields and also, when inseminated by a 'beef' bull, produce steer calves which grow quickly and give the best quality of lean meat. About seventy per cent of all cattle in Britain are now Friesian or Friesian cross-breeds, although the beautiful and delicate Guernsey and Jersey cows are still kept for the high proportion of butter fat in their milk. Of the beef sires over sixty per cent are Herefords and all but about five per cent of the rest are our famous Black Aberdeen Angus bulls (a breed whose origins have been traced back to 1523), or the heavy Charolais which have come over from France. The big, white face on the black and white body of the Hereford × Friesian steer is therefore the one very likely to gaze curiously at you over the hedge.

A good beef animal is rather rectangular, seen either from above or from the side, with a short neck and stocky body, giving the smallest proportion of the poorer cuts. A good dairy cow, by contrast, is more wedge-shaped, from either viewpoint, with more delicate head, neck and shoulders and more rounded thighs. The modern dairy cow of course can give a great deal more milk than a single calf needs, although immediately after birth a calf will drink a daily gallon of milk and will be taking 1½ gallons when it is a fortnight old. The cow can rear several calves, but where the product is milk, the calf is hand fed and the cow remains in the dairy herd.

The calf will take milk from a bucket, at first sucking the cowman's finger, and this demands both skill and patience. Too much milk will cause 'scouring' (diarrhoea) and this can be catching, so that it is wise to rear calves in individual pens with a good, dry floor in a warm building, and each should be examined at least twice a day. After about a week it will be eating a small quantity of solids, such as flaked maize, crushed oats or barley, or linseed cake and should have its vitamins. At about six weeks the calf will join a group of twenty to forty others, in a pen with water constantly available, eating concentrates and hay, which gets it accustomed to the grass it will graze from the meadow.

The good heifer calf will stay on a dairy farm until she is old enough (at eighteen months or even less) to be inseminated and then to graduate to the dairy herd, but the steer calf born in the spring may be sold in the autumn to beef fatteners and

13 *Beef calves housed in cubicles for the winter with their dams*

then may be 'finished' for slaughter at between fifteen and twenty-two months, having lived at least one full season out on the grass, and at least one winter in the yard on hay and cereals. Some farms both rear and fatten beef cattle, but often the fattening pastures are not suitable for rearing and winter fattening may take place in yards or cattle houses on arable farms – largely on the farms' own produce – which in summer have no fields to spare for cattle of any sort. These 'stores' or cattle for fattening are also bought in from the beef (as distinct from dairy) herds in which they have run with suckling cows, often Friesian × Hereford cows mated with a 'beef' bull and they come from upland farms or from Ireland. A great deal of our beef comes from Scotland and will have been slaughtered at 2 years or 2½ years at the oldest, although the animal is not fully grown until three years old. This is not only because our taste is for the younger and tastier meat, but because there is greater demand for the smaller joints. And of course, there is beef or 'beef products' from cows which drop out of the dairy herd.

All successful systems of stock rearing are essentially simple and the cheaper systems are necessarily preferable to the dearer ones. But there is no substitute for a good stockman and nowadays he has a whole range of daily tasks. Stockmen are hard to find and this is not surprising, because livestock have to be cared for seven days a week and every week of the year – and where there is stockbreeding the cowman or shepherd, just like the general practitioner called to a maternity case, is likely to have his night's sleep interrupted. On a large farm, with careful rostering, the stockman will get his weekly day off, but the junior may still find that even his Sunday is not free of some essential feeding, and on the small farm, if the stockman is in his good suit and out with his wife, it is the farmer himself who is feeding the stock. The head cowman on a farm with a large dairy herd may well be paid the same as the manager who has a degree in agriculture (and the manager has no fixed limits to his working day).

The attitude of the stockman to his animals is one which may seem strange to townsfolk. There is no margin in farming for sentimentality. The shepherd may go out in the bitterest weather to find and feed snowbound ewes on the Yorkshire moors, but livestock are seen as purveyors of meat, milk, eggs and wool; the animal which does not do well will be culled from the herd and sent for slaughter, and the slaughter-house is the final destination of the whole herd at the right economic moment. And yet the stockman or the shepherd is more than observant and efficient with his animals. He is quiet and gentle and takes pride in their condition. Instinctively he recognises as a countryman that the brute creation has its rights alongside the human, and suffers as we suffer and has the same claim to be spared unnecessary suffering. In the big city the human being sees only other human beings, generally concentrated in excessive numbers, with pets that are no more than adjuncts of their owners, whereas the country boy has grown up in a world which he shares with other animals, each with its own independent way of life. The stockman takes it for granted that his animals will not live beyond their profitable term, which is very much shorter than their natural term, but he is fond of them and respects their needs.

Is there any more satisfying sight than contented cattle on a rich pasture, or any sight which seems more typically British? The delicate fawn, with white markings, of the Guernsey cow or heifer, recognisable even in a herd seen at a distance, the lazy swish of the tails, the clump of old trees planted for shade: this is the sort of picture anyone can recapture who has travelled through the countryside. In the evening too, one may feel the cool air rising from a pasture, distinctly hear the grass being torn and the long, slow snuffle of cattle whose forms are no more than darker patches in the dusk. I remember one fine Ryegrass–White Clover pasture in late summer, Friesian dams with crossbred Charolais calves at heel, strangely grey-speckled creatures some of them, some fast asleep in the grass, the cows grazing, ambling gently across the field, or quietly ruminating, quiet and content over the whole sunlit scene. Simply to contemplate it for half an hour seemed to unwind all the day's tensions. It looks so beautiful, so peaceful, so easy. This is deceptive: one forgets all the skill, the care and the work which lie behind this scene, but the beauty and the peace are real enough.

4

MIXED FARMS AND FARMING MIXES

After the ploughed land with its crops and the grassland with its herds, the next thing is to build up a picture of the many different ways in which the two are combined in general farms, of how the rotations work out in practice, of how the different sorts of farming pay and of what keeps a man, and his wife, on the land. True, sheep, pigs and poultry are waiting for the next two chapters. They are vital to some farmers and to the country: wool was Britain's first major export in medieval times and sheep have left the imprint of their golden hoof in the lovely old buildings still standing in parts of the country where sheep are now rarely seen, for the onward march of the plough has driven the herds of sheep mainly to the hill and upland farms. Pigs and poultry in the main are now indoors in 'factory farms' and although these enterprises may be part of a general farm, they are a specialised type of farming.

The remaining speciality, before we turn to the varieties of mixed farming, is vegetable growing. This book chooses not to regard fruit growing, whether soft fruit or apple, pear and plum orchards, as farming. And this despite the attractive picture they present, with trays of shining, round apples in autumn; the sweet smell of the blue smoke drifting up from bonfires when the trees are pruned, for they are skilfully kept down to convenient picking height; despite the glory of the spring blossom and despite the peculiar charm of the toy tractors which are built to edge between the trees. Fruit growers are unfortunately even more at the mercy of the weather than market gardeners, but today a large proportion of the fruit we grow at home is better presented than what we import and its quality is always better.

Nearer to our concern are the great tulip fields around Spalding in Lincolnshire – flat sheets of colour apparently gone mad, until one comes across a field with all the colour stripped away into a vivid pile of torn off flower heads by the ditch at the field's edge and realises that the crop here is not the flower, but the bulb. What *is* our concern, with techniques akin to more general farming, are the areas mainly occupied by vegetables, on deep alluvial loam by the Wash, around Evesham, on the moss soils of Lancashire around Ormskirk, and in Scotland in the Lothians, Angus and Perthshire. Some of the villages out in the Lincolnshire Fens seem as inward-looking and cut off from the outside world as the older mining villages. There is an end-of-the-world atmosphere about the flat, treeless fields, divided by deep dykes and straight roads which run on causeways up above the fields, and at the end of the journey are the great embankments built to hold back the sea. Here much of the work is done with narrow-wheeled, row-crop tractors, with tool-bars set either in front or between the front and back wheels, as well as behind – hoes, harrows, rotary hoes, 'down-the-row' thinners, transplanters for brassica crops are the sort of tools used. This is close, intensive cultivation with sometimes as many as four and often at least one or two men intently at work on the machine behind the tractor driver, as he edges

14 *Transplanting Brussels sprouts – a cold and laborious business*

slowly and precisely across the field.

All this work is only justified when the soil is deep, rich and in good heart and may yield two or more crops a year. But these are perishable crops, susceptible to the weather and, unlike dried grain, cannot be stored until they fetch a good price. On the contrary, they must be washed, graded, sorted, packed and sold soon after they are taken off the field. The grower must know exactly what is required: the smallest carrots go whole into tins; the very large ones are diced for soups or dehydrated; the medium sized go to the Midlands or London, the larger ones to to the North. A big buyer will order a regular supply of uniform, graded quality, which is often most easily provided by a co-operative of small producers, but the market is easily glutted and the grower may get as little as a quarter of what the housewife pays. What is certain is the skill and the hard work; much less certain is the reward at the end of the day.

Of course chemicals, which are not cheap, can nowadays take the place of hoeing, and flat land, divided into long strips (almost as in medieval open fields) can be worked by machines. Nevertheless, on an arable farm, one man may plough all day without another worker in sight, whereas spring cabbage, peas, runner beans, Brussels sprouts and cauliflower inevitably involve labour. The crops are all the vegetables we buy: carrots (from deep, light, sandy soils), onions, beetroot, lettuce, broccoli, early potatoes. Some villages grow a sequence of vegetables, transplanting seedlings from seed beds to mature in the fields from which the first crop has been stripped. Other farmers grow one vegetable and follow on with ordinary farm crops, but only those devoted to vegetables really understand the grading and marketing of vegetables other than potatoes.

Potatoes appear with sugar beet as cash crops on many

farms and both clean the soil and add to their income, but on these farms cereals have first place. On some good soils there will be a rotation of potatoes, wheat, barley and sugar beet, but the beet now predominates in Norfolk and Suffolk and the potatoes in Lincolnshire and around the Wash. High cereal prices, in a world from which grain surpluses have disappeared, and the disappearance of so many farm workers, have led some farmers to grow nothing but cereals, using only artificial fertilisers. These are the farms (most remote from our story-book conceptions) with no animals, given over to large machines, where every possible process is mechanised, where 600 acres may be worked by a couple of men in place of the fifteen or twenty who were there before the war – and where the farmer is risking deterioration of the soil structure for lack of humus. Even so, a £10,000 combine, which he must have, will only be used for a few weeks in the year.

Come now to the rich soils of the Fens, flat, black, treeless, with horizons of the sort familiar to the sailor, and 'drains' or rivers banked up to flow high above the fields. Or those rich, flat fields in the Trent estuary, which give an uncanny impression of belonging more to the sea than the land, or parts of Essex, the Lothians in Scotland, north-west Lancashire, the Isle of Thanet in the Thames, or some warm coastal areas in Devon, Cornwall or across in Pembrokeshire: here are rich and remunerative farms, growing 'high return' crops such as potatoes, sugar beet, celery, Brussels sprouts or peas (for humans, not for stock-feeding). But potatoes and sugar beet call for labour at the same time and a careful balance must be kept, because profits fall quickly if labour and the new heavy machinery are not evenly employed throughout the year. One idea is to grow annually the same acreage of each crop, but simply to change fields.

The 'cash roots', sugar beet and potatoes, appear too on those arable farms which combine cereals with dairy herds. This is a judicious balance. The livestock can be fed off the farm's own barley and oats; they will eat some of the unsaleable products of the ploughland, such as 'tail' corn or 'chat' potatoes (both too small to sell); they will turn the farm's straw into good farmyard manure; and the grass leys they will graze make a very wise break in the cereal-roots rotation. Moreover, the wet July or August, which may spoil some of the cereal crops, will at least offer some compensation by prolonging the period in which the grass continues to provide good grazing. Or of course there is the farm which grows only grass, and on the grass feeds its dairy herds, or combines them with fattening beef stores, and perhaps has a few acres under the plough as well. There is a rich variety of farming in Britain.

The sort of farm which appears in the story-books we read as children and in the old nursery rhymes, is the small mixed enterprise, with the traditional bit of everything. This is the sort of farming which has been typical in most parts of the country for the last three centuries, with some permanent pasture and some leys down to grass for three to five years, and the remainder, up to half the holding, given over to cereals, roots to see the livestock through the winter, and a few acres of potatoes or sugar beet. This mixed farming evens out the demand for labour throughout the year; spreads the risk if one crop fails, and spreads the cash flow more evenly over the year. But there is not much in the way of high-yielding cash crops and the enterprise does not really support a modern standard of living unless it is a farm of several hundred acres. If there are no other sources of income, the farmer and his son will be working all the hours the daylight gives them and still be hard pressed to put by the money for a new tractor when the old one wears out. It will be a simple life, a man's life all right and many still seek it, but a life which finds its satisfaction in the work, because it contains little else.

On the mixed farm, the weakest link in cost accounting may be the return from the large quantities of home-grown cereals fed to livestock and not properly recorded. Low yields from livestock can be checked from sales records and excessive feeding from feed records, but margins are so small that it is easy to lose money if each grazing field does not support as many animals as it should. The correct method of avoiding this is to use a calculation involving 'livestock units'. These convert all grazing animals to a common unit, equivalent to a dairy cow of medium yield. You convert all your grazing stock into these units and divide the whole forage acreage by this total, which gives the number of forage acres per livestock unit. The answer (on an ordinary lowland farm something like 1½ acres per unit) will tell you how you compare with what is normal for the district. It is a useful calculation for a farm large enough to run at least something like a proper office, and a survey published in 1974 emphasised the key role of good management: some dairy farms obtained a yearly 600 gallons of milk per acre, against an average of 247 gallons. But precise calculations are less readily made by a man who has been outdoors, whatever the weather, working with his own hands since seven in the morning. He will know however, almost subconsciously, that an average yield per acre of wheat is now about 35 cwt, of barley and oats around 30 cwt, of potatoes 11 tons and of sugar beet 16½ tons – an increase over about 12½ tons a dozen years ago.

The thing to remember about beef is that, except on a minority of farms in this country, it is a sideline, a by-product of dairy farming, in the sense that three-quarters of the home fattened beef is from calves born to dairy cows. Although home beef production is up fifty per cent on what it was before the war (when we relied so heavily on Argentina), producing beef meat is a biologically inefficient use of feedstuffs and economically a questionable use of our scarce land.

Herds of suckler cows with beef calves wander freely over relatively large pastures. Because the cost of special breeding stock and of feeding over a fairly long period amounts to a considerable investment, while profits are problematical, a good deal of land on mixed farms that was previously grazed has now gone under the plough. And this despite the fact that on beef farms the cows will calve outdoors and that the cattle will not come indoors until the calves are at least three or four weeks old. This reduces the risk of infection and when they come indoors it will only be into 'cow kennels', quite cheap accommodation made up with bales of straw, corrugated iron sheeting and some poles, perhaps with a cosier area into which the calves can creep and receive small quantities of concentrated feeds aways from the cows. Again the constant theme: there is no margin to spare on housing, food or labour, and crossbreed cows (probably Friesian crossed with a 'beef' bull) are preferred to pure breeds because they live longer, conceive more regularly and make better mothers. On heavy land they must come indoors in winter, because of the damage their four hooves inflict on the fields in wet weather, killing the sward mixtures which they find most palatable and encouraging buttercups and daisies. Hence the bitter remark, echoing down the centuries from our peasant forebears, 'a beast has five mouths in winter'.

Most of our Midland farms, famous for their permanent pastures, were all grass before the war, with no tilled land – a form of what is called dog-and-stick farming, because no other implements were used. Now the majority have a considerable area of ploughed land and of 'up-and-down' or 'alternate' farming; that is, fields which will be ploughed for some years and then form leys for a period before being ploughed up again. The effect of post war changes on our meat ration is what you would expect: beef eating down since the war by about one-third (although more produced at home), mutton and lamb halved, the various forms of pig meat more than doubled, poultry meat trebled. Hence the importance of 'factory farming', which produces pigs and poultry. They can be reared intensively. As one writer put it, this is like adding a second storey to the farm and good land is becoming too precious to devote to stock rearing.

The formal rotations, to which the great eighteenth century pioneers devoted so much thought and enthusiasm, are now less rigid. The common form on mixed dairy-arable farms is to grow long leys around the milking parlours, with meadows for hay and silage perhaps elsewhere, and more regular rotations on the arable fields. A four-year ley may provide an ideal break between cereal and root crops, breaking the sequence which invites pests and diseases and building up soil fertility. A common rotation nowadays would be wheat–barley or oats–sugar beet–barley–a one year ley (perhaps for hay)–potatoes–wheat. Or on land unsuited to roots the rotation might be: wheat–wheat–barley, followed by a long ley. If the farmer grows turnips or swedes, they are the principal break between cereal crops.

'Catch crops' can sometimes be snatched between two of the main crops of a rotation. For instance, an enterprising farm manager may simply put a heavy cultivator through the autumn corn stubbles and then broadcast Italian Ryegrass, which will give his cattle welcome early spring grazing before he sows another regular crop. Because wheat grows a strong straw, it successfully follows potatoes or other well-manured roots, where other cereals would react to the rich diet by making excessive lanky growth and keeling over in bad weather under the weight of the ripening ear. But in all his rotations, the farmer must remember that land ploughed, sown and harrowed down to winter cereals will lack what it must be given after a few years, namely the opportunity for deeply ploughed and well-turned furrows to lie and mellow through the winter under the alternate wet and dry, frost and winter sun, untidy, but preparing a fine tilth for the spring.

Farm workers are poorly paid for their multiple skills and increasing responsibilities, for their willingness when necessary to work all hours in all weather and for the fact that they are tied to their jobs, in the sense that they must be available whenever the state of the soil or the ripeness of the crop demands them. The agricultural worker does not strike, but he votes with his feet; sons do not succeed fathers and the total number of full-time employed men on the land has fallen from nearly 600,000 in 1952 to less than one-third of that number now. Those who remain each do more, with more machinery and equipment, more ready-processed feeds and modern planned buildings – the number of dairy cows for which one man is responsible has multiplied tenfold in thirty years. But the reduction in the number directly working on the land is deceptive, because farming now employs and is supported by quite a large section of industry. Our society still gets its values wrong, rewarding the manipulators a hundredfold more than the makers and the growers, but the men on the land can no longer be worked until they are prematurely worn out, like ill-used animals, as they often were right into the 1930s. And with the departure of the horses, although their ghosts still linger, only the stockman still needs to live on or very near the farm.

It is still a job which attracts more trained applicants for posts as farm managers than are needed, so that many of the young men graduating from agricultural colleges go into the advisory services, research or teaching. And there are still more men wanting to farm than there are farms to buy or lease. Growing food, working out of doors, still attracts the man who could not see himself at an office desk. But he must find a wife with no longing for the bright lights, who understands that he will be tied to the seasonal needs of the farm and will virtually get out of bed into his gum-boots, yet who reckons it ample compensation to bring up her children where they can feel the rhythm of the changing seasons. Mr James Merridew, Director of Nafferton Farm connected with Newcastle University's School of Agriculture, is probably right in claiming: 'For the greater part of my life it has needed a keener brain to make a living from farming than from any other industry'. Partly this is because, when there is sufficient of a food, then an increase in supply does not stimulate an increased demand, but simply gluts the market and reduces the price.

The Common Agricultural Policy of the European Economic Community seeks to counter this by a tariff wall to hold back imports, while buying up and removing from the market any supplies which cause the price of a commodity to fall below a predetermined level. This has kept up the prices in the shops and at the same time encouraged the farmers on the Continent to continue producing far in excess of the quantities bought by the public at the artificially maintained prices. Ultimately vast quantities particularly of butter and beef, removed from the market so as to sustain their prices, had to be disposed of deviously or outside Europe. As a system, this does scant justice to the intelligence of politicians, but the not ignoble intention was to make less painful and more gradual the necessary process of easing small-holding peasant farmers off the land in France and southern Germany. In the complex arrangements between countries, the main cost has fallen on German industry, which has borne it easily. (In the eight years 1962–70 Germany paid out 804 million 'units of account' and France received 914 million units.) Also one must remember that the politicians were faced by violent agitation, with small farmers blocking main roads, hanging pigs from city lamp posts and introducing a cow into the council chambers of the EEC. More important perhaps: in France farmers and their workers make up thirteen per cent of the working, and voting, population. Our corresponding figure is three per cent, and they can never swing an election.

The British system of using taxpayers' money to make up the farmer's income sustained home food production without increasing shop prices above the world market levels to which we were accustomed. At the same time the Annual Reviews made it possible to encourage or discourage the growing of particular foods with a flexibility which eludes the Common Agricultural Policy. British farming responded with the achievement of increasing efficiency and avoided the disastrous times which followed after the effort of the 1914–18 war. The achievement in this period is set out in figures: before 1939 we produced only 23 per cent of our wheat and flour, in 1972 60 per cent (we cannot in any case grow the hard grain needed for long-keeping bread and this comes from North America). Of our barley we used to grow under half and now grow practically all; in poultry and pork (as distinct from bacon and ham) we are self-providing, where before the war we imported about one-fifth; with eggs the increase to self-sufficiency has been twice as great. We produce all our own liquid milk and 60 per cent of our cheese, but only 22 per cent of our butter. Of beef and veal 85 per cent came from home stores in 1974. We are very fortunate too in having only half as many small holdings as in the EEC countries. These are the holdings of under twenty-five acres, comprising nearly two-thirds of the total number of farms in the EEC and in today's conditions no longer viable. The tendency here too is towards larger farms and the small farm can only pay if it produces some speciality with a very high yield per acre.

In recent years Britain's agricultural productivity has been increasing at about twice the rate achieved by industry. The National Farmers' Union maintained that in 1974 one man on the land was feeding forty-two townspeople, compared to between twenty and twenty-five in France, West Germany, Italy and Eire. But what of the future? In 1972 two well-informed writers on agriculture commented that 'there are stock piles of grains, wool and butter, but one never hears of beef being stored because there is a glutted market and this situation is likely to continue for a very long time'. Two years later a spokesman for the Farmers' Union of Wales was quoted in the press on 'over-production on a massive scale, with butchers being offered more beef than they can sell to the housewives. Prices paid to farmers for their animals have slumped in a most dramatic manner.' In the face of this, it would be a rash man who ventures on prophecy! But in recent years North American stockpiles of grain have disappeared. In a hungry world, with its remorselessly growing population, it is unlikely that cereals and feed grains will become cheap again, while the oil stranglehold keeps up the price of fertilisers and fuel. Land and machinery prices both doubled in the decade between 1960 and 1970, *before* the onset of rapid inflation.

It seems improbable then that we shall see cheap food again and one only hopes that the farmers will be saved what I think has been called the 'pig cycle', but which also occurs with beef and poultry. This is due to excess supplies reducing selling prices below cost, so that farmers slaughter their breeding cows, sows and hens, which later leads to such a shortage that prices shoot up, which attracts too many farmers into meat production – and so on. The trouble is compounded by the long time which elapses between insemination and the offer of finished beef animals for slaughter, although with the sow the period from pregnancy to pork is shorter.

Another long-term trend is the great increase in the working capital needed to run a farm, and this capital now quite often comes from investment by institutions in place of the traditional landowner. He, in order to increase the return on his capital, now inclines more often to farm the land himself, instead of leasing it to tenant farmers. Another sign of the same pressures is that the very big pig and poultry 'factory farms' tend increasingly to be owned by the large groups which manufacture the concentrated feeds on the one hand and on the other distribute the produce through their supermarkets. More sophisticated marketing of crops, perhaps through farmers' co-operatives, and ever more careful grading are also likely.

Contract work will also probably increase. This is not new: in the brief heyday of the steam engine on the farm, ploughing by two of these monsters, hauling a multi-furrow plough to and fro across the field between them on a wire hawser was not uncommon and this was done by contract. Until after the Second World War, the steam engine used to arrive in the stackyard towing the threshing machine and this was also outside contractors' tackle. Now some ploughing, draining and aerial spraying are done under contract. Also many crops, such as Brussels sprouts, peas, beans, carrots, raspberries, potatoes for chips, crisps or canning, are grown under contract with the freezer, canner or packer, generally a large firm with more capital than most farmers. Sometimes the farmer may be given precise instructions by the processor, so that he becomes almost an employee. Sometimes the processor may even hire the field, knowing that he has good, clean land, and actually carry out the whole operation himself, from seed to harvest. Or the farmer may grow the crop, but the contractor harvest it, dovetailing the processing of the vegetable with that of his other foods, such as fish.

All these arrangements give the contractor precise control over the quality of the crop, the speed of harvesting, essential in the case of frozen foods, and the opportunity to deploy specialised equipment beyond the means of the farmer. Peas for freezing are a particularly good example. The factory determines the time for harvesting to secure a uniform product, taking readings with a 'tenderometer'. Readings of 95 to 105 are suited to quick freezing, but up to 120 (which implies a higher yield per acre) is acceptable for canned peas. The field must be fairly near the factory, so that the peas are frozen almost as soon as they are picked. On the appointed day two or three 'viners' lumber into the field in the freshness of the early morning, accompanied by several lorries to ferry loads quickly back to base. A mobile workshop parks by the gate in case of breakdown, with a car or two for quick communications perhaps £80,000 worth of equipment deployed in an almost military operation. We came across one such field in Lincolnshire in the evening, as the viners were just finishing their task. The lorries had left; the field was stripped; the dusty great machines and the tired men prepared to reform convoy. They moved off into the evening with the mobile workshop in the rear.

5

UPLAND FARMS AND THEIR SHEEP

In the past 'the sheepfold was the sheet-anchor and mainstay of husbandry. The same flocks that were driven each morning to feed on the downs, wolds, hills, heaths, commons, and pastures, were brought back to the ploughland at night to be folded behind hurdles, in small and crowded pens, so that they would deposit there the residues resulting from their day's nibbling, chewing and drinking. The sheep were thus used as four-footed muckspreaders, or, rather, as mobile combinations of fertiliser manufacturers, distributors and spreaders, fetching their own raw materials and processing them, and delivering and applying their products. . . . The whole farm was laid out and run to suit the sheep, but only on the strict understanding that they devoted themselves to fertilising the land for corn and grain.' (Eric Kerridge, *The Farmers of Old England.*)

But our crops no longer depend on this manure and sheep have almost vanished from our lowland farms, where once their wool was the farmer's golden fleece. In the whole of Britain, the only large area of lowland grass now populated by great flocks of sheep is Romney Marsh. The plough has captured much of the uplands and even where they are still grazed, sheep now give way increasingly to cattle, which pay better. The two can share the same fields and there are possibilities for improving upland grass, even for more intensive and more profitable sheep farming on some lowland farms, but sheep have not kept pace with the great changes which the last thirty years have produced in cereals and with cattle. There have been few major advances in either the management of sheep or in their breeds in the last generation. Sadly, sheep farming has been called 'an industry which retains a greater element of tradition than almost any other major branch of farming'.

It has been forced to the mountains, the moorlands and the hill sheep farms, where the land and the climate offer only the growing of trees as an alternative to the growing of lambs. And in going up to the hills, the sheep go back to an even older tradition than the Middle Ages – that of Biblical times. As Isaiah wrote, 'their pastures shall be in all high places' and the imagery of the sheep and their shepherd is woven through the whole texture of the Bible. This was a pastoral people, living on the uplands. 'Their stage slopes away from them, every apparition is described as *coming up* . . . Israel *treadeth upon his high places*, as if mountain tops were a common road; *the Lord marcheth upon His high places*, as if it were a usual thing to see clouds below, and yet on the tops of hills'. This, from George Adam Smith's classic *Historical Geography of the Holy Land*, sets the scene for most of our sheep farming today.

You see them on mountain and moorland, the Scotch Blackface, one of the most picturesque sheep in the world, the Herdwick in the Lake District, the Swaledale from the northern Pennines: wild open spaces, long empty views stretching away at their feet. George Henderson describes them in a harsh winter: 'the sheep came drifting along like bundles of cotton-wool as their fleeces trailed in the snow'. They graze enormous often unfenced acreages, nibbling the spare grass and the heather. I remember the Swaledales on a wet day when the high Yorkshire Moors were obscured in mist, the only sound the constant trickle of water, as the ewes with their lambs materialised out of the mist, moving sure footed, silent on the steep slopes, picking a living where there seemed to be none, disappearing into the mist again within a few yards.

The lambs are brought down from the harsher climate in the autumn. The wether lambs will be sold for fattening on lower land and the ewe lambs destined to replenish the flock will also overwinter in easier conditions. The actual lamb 'crop' on these high farms depends on the weather. Although lambing waits until April or May, lambs may still be lost through exposure. A good season may yield no more than eight lambs to ten ewes and a severe spring may halve this. In catastrophic winters like those of 1947, 1963 and 1968, the breeding ewes themselves were decimated.

Perhaps it is the stoic self-reliance and self-sufficiency of mountain sheep which explain their emotional hold on those who have shepherded them and call them the most intelligent of our livestock. George Henderson, long after he was successful on his mixed farm, still maintained that 'the finest sight I know is a great flock grazing on a green mountainside' (echoing the Psalmist's 'the pastures are clothed with flocks'). Beatrix Potter, whose children's books are still read by generations born since her death, ended her career as an acknowledged authority on Herdwicks. She learned to admire their endurance when overtaken by blizzards on the fells. Buried in the drifts, they would survive under the frozen vault of the snow, for a week or a fortnight, nibbling patiently at the moss and their own wool until they were found, 'dogs scratching, and shepherds prodding the drift with the long handles of their crooks'. She loved the sturdy, small, smoky-blue sheep, tiny hooved but with strong, thick legs and the rams with fine curling horns. Sheep dogs too seemed to her to inherit their skills. That 'quick dark shadow', which will hold a flock while the shepherd inspects them, will sort out a flock, holding back one sheep and letting go another, rounding up stragglers, herding the sheep on a hillside which has neither hedges nor walls.

These active mountain sheep grow a comparatively heavy fleece and from the mountains come most of our breeding ewes. After three seasons on the high places they will be sold to a farmer down on the lower land, who will have one or two more lambs from them in easier conditions. On hill sheep farms, where conditions are less severe, the 'yield' of lambs is higher; more of the grazing land is enclosed, which makes for better control and it is easier to supplement the winter ration. But with more intensive stocking it becomes more important that the growth of the pasture and the appetite of the flock should move in pace with each other. As with cattle, when a flock gets restive, spending more time looking for food than nibbling and cudding, then it is ready to move to a fresh pasture. The art is to keep the grass short and leafy, for lambs do no good on rank grass.

The upland farmer (we are now below 900 feet) will buy the hill farmer's four- or five-year-old ewes – breeds such as the Cheviot, Exmoor or Kerry Hill – and will take one or two lambs from them, again under easier conditions, before they are mated with a 'longwool' ram. The Border Leicester for instance, mated with a Cheviot ewe, produces the Scottish Half-Bred, a famous 'grassland' half-breed ewe, prolific, a good mother, whose lambs fatten readily. At this level the fields will be enclosed and it is possible to be more ambitious about gradually improving the pasture. Ploughing and reseeding may be too

15 *Sheep grazing a good paddock*

expensive; may face the farmer with huge numbers of exposed boulders to be disposed of and may also upset the natural drainage. But lime, slag, and carefully controlled stocking will do a lot and you will see that in places where sheep regularly shelter, their own droppings improve the sward. It is possible to turn the soil with a rotary hoe and then to sow quickly-established Perenniel Ryegrass and White Clover. This will increase the area from which the grass can be preserved for the winter feed and on the uplands winter feeding remains the limiting restraint, as it has been since medieval times.

But for all the romance of the famous breeds, for all the bustle and stir of the autumn sales, as the grassland improves and the altitude diminishes, the sheep compete less successfully with the plough and with cattle. There are many hundreds of thousands of acres of former moor which now carry good grasses and of land now alternately ploughed and in grass, which twenty-five years ago lay untouched. This results from a long-sustained post-war policy of 'farm improvement' grants and it brought beef growing herds to the marginal farms. Since 1950 this has given us perhaps an additional quarter of a million breeding cows – and helped us to replace the beef we used to import. Sheep and cattle can be complementary, but 100 suckling cows are no more than a part-time job for the stockman. Their economic equivalent is 600 breeding ewes and they occupy the shepherd fully. Yet sheep will finish a pasture after cattle, for cattle eat out the top-growing grasses and this encourages the finer grasses favoured by sheep – cattle pull where sheep nibble.

Sheep too will act as winter scavengers on a field after cattle have gone into the yards. The breeding ewe is under stress only for the last two months of pregnancy and during the four months in which she suckles her lamb, whereas the suckling cow is physiologically stressed for at least ten months. Where the grass is only growing for six months of the year, the ewe can still find enough for herself when the cow already needs supplementary feeding. Above all, the intestinal worm parasites of sheep find no host in cattle, so that sheep prosper if they follow cattle on to a clean field.

So much is still traditional: the shepherd with his dog, perhaps on a motor cycle but perhaps still on a pony; the lonely stone built farmhouse tucked low into the shelter of the hill-side; the dry-stone walls running up steep fields; the patient sheep, as I once saw them all over an upland village, their inborn wisdom finding them shelter from a sudden hailstorm in every hollow, behind every boulder and the walls of cottage gardens.

16 *Ewes spending the winter in the protection of cheap buildings*

But experiments and changes do take place, such as keeping breeding ewes indoors on lowland farms in winter, especially on heavy land, where they would spoil the fields by treading them when wet. They need only cheap buildings, and the cost of these can be set against the cost of carting winter feed to them out of doors. Morever, not only the ewe, but also the shepherd is under less stress if lambing takes place indoors. More lambs also survive, at least one for each ewe, and their value too can be set against the cost of buildings. When they go out into the spring fields, this new, more intensive system moves the flock in succession through six or eight paddocks, changing every three or four days according to the growth of the grass. Sometimes too a farmer buys in a 'flying flock' and sells it again within the year, either ewe lambs for subsequent sale as young breeding ewes, or store lambs for fattening.

All these are attempts to make sheep pay better and in-wintering is certainly becoming more common. The crop is young animals. The ideal is for every ewe to be in conditions in which she will bring up at least one and preferably more than one lamb every year, say, 120 lambs from 100 ewes. The trade in fat lambs starts at Easter and fat lambs 'come off their mothers' throughout the summer. As the summer advances, so prices usually fall and the lamb which is sold even a fortnight earlier than another may fetch a considerably better price. Hoggets are fattened during the autumn and winter, or kept over the winter and fattened on the following summer's pasture. Then there is the wool. Ewes will give 3½ to 4½ lb, wethers up to 7 lb. The long-wool breeds yield 12 to 16 lb and a well fed ram even more. Some famous breeds we have named, but 'there are some forty recognised breeds of sheep in Britain and a large number of crosses as well as distinct types within some of these breeds and crosses. The farmer who first comes into fat lamb production must have some of the feelings of a dog let loose in a butcher's shop.' (Cooper and Thomas, *Profitable Sheep Farming.*)

'I'm never more than 200 yards away from the flock. A shepherd is always close – he always has been, hasn't he?' This is Ronald Blythe's shepherd in his famous book about a Suffolk village, *Akenfield.* The shepherd's year begins in the autumn, when those ewes who are fit to breed again are sorted out from the flock and sufficient young ewes are added to make up the full number. Generally these new recruits will be mated so that they first lamb when they are two years old. Before mating their feed improves. This may reduce the number who fail to conceive and may increase the number of twins (although of course on moor and mountain one lamb is all a ewe can manage). It will also bring the ewes into season more uniformly, so that the lambing season will be over more quickly.

The ewe is on heat for twelve to twenty-four hours and it is not uncommon to employ as 'teasers' rams which have undergone vasectomy, so that they are sexually active but impotent. The shepherd in *Akenfield* continues: 'The tups go into the ewes about the first week in September. These aren't the proper tups; They just bring the ewes on. We call them teasers. It means that when you put in the proper tups the ewes are good and ready for them, and can be served in one bunch. Let them love together and they'll lamb together, and that will be convenient. A good ram will serve fifty ewes after the teasers have been with them for a month. I work it like this. I put half my proper tups in with the ewes for two days, then take them out and give them a rest while the other half have a go. Each ram has a harness full of crayon strapped round him, so that when he jumps he marks the ewe. I then know how many ewes are coming. I change the crayon – the raddle – every fifteen days.

So the first raddle will be blue, then red. In the olden days they painted the jumped ewes with red ochre but now we have this system of telling. If all the ewes are covered the first fifteen days and none of them come back, then I take the tups out. I leave them roughly three periods to come over – about forty-five days all told.'

When the ewes are pregnant they should not be too well fed until about the last two months of the 21 weeks pregnancy, when they will probably get some concentrated feedstuffs. This may continue for some time after lambing if need be, in order to maintain the mother's flow of milk, since plenty of milk is vital to the lamb in its first two months. Lambing out of doors is still usual and the ewes move to the lambing field about ten days beforehand, generally in March or April. Small pens are built for ewes newly lambed and larger ones for those about to lamb or with the lambs already at heel. This, especially in harsh surroundings, enables the shepherd to see that all the new born lambs are properly mothered and he will not leave the flock by day or by night. It is rare that some lambs do not need the shepherd's help and if a ewe produces triplets, or if he loses a ewe, he must find a foster mother. And it is true that if a ewe has lost her own lamb she can often be induced to suckle a strange lamb if it is covered with the skin of her own.

Except on mountain or moorland, the young lambs' tails are docked and the ram lambs will be castrated. After two or three weeks (as in calves) the rumen begins to develop rapidly, so that the lamb can digest grass. Weaning – generally after four months – simply means taking the ewes away from the lambs. And it is wise to take them out of earshot of the plaintive bleating of their young. They move on to poorer pasture, which more rapidly stems the flow of milk.

In the south and on low land the wool is sheared about midsummer, but in more rigorous places not until July or August. It is the warm weather which causes grease to 'rise' in the fleece and until this happens shearing is more difficult. For the most part, lambs keep their wool until their second summer. A fortnight after shearing, the sheep are dipped. This was compulsory in order to control scab, but the parasite causing scab was eradicated by 1952, although unhappily it reappeared in 1973. Nevertheless, the maggot fly, ked, sheep tick and sheep louse all remain to plague the flock, unless they are either dipped or, nowadays, sprayed. This is the last of the shepherd's major routine yearly jobs, but he is always looking out for footrot and sheep are prone both to lung worms and to two different sorts of intestinal worms. Their health will depend on his awareness of their need for periodic 'drenching' against these and on the management of their pasture. Lambs, above all, need fresh grass not recently grazed by ewes, for it is unhappily sheep which infect sheep, especially when they graze close packed on the pasture.

In all this, you will notice the limited role of the male animal. The bull is now rarely on the farm, which makes for safety. The boars and rams are few and appear briefly. The steer, the hog and the wether live only long enough to put on weight. Only the breeding ewe, the sow and the cow have a longer, productive life. It is the *ewe*-lamb which since Biblical times has symbolised one's most prized possession.

6

INDOOR FARMING

Pigs, contrary to common belief, are clean, active and intelligent animals. In my view, there are no other farm animals which so visibly enjoy themselves. Piglets do not skip and gambol with that charm which has so endeared Easter lambs to poet and painter alike, but they do have charm and are more purposeful than lambs, especially where food is concerned. Pigs positively enjoy digging up pasture and woodland with their strong hoe-shaped snouts and if they are on good grass it will satisfy a fair amount of their need for bulk food. Indeed, they can enjoy fresh grass clippings if fed direct from the mower, but they are not ruminants. Like us, they possess only a single stomach and cannot live on grass. What they seek with such vigour when they run loose on a pasture are succulent morsels, the minerals they need in their diet. Their active snouts and sharp hooves will soon destroy the pasture and, unless it is free draining land, turn it into mud, although this can be avoided by putting copper rings through their noses so that they cannot be used for digging.

Few farmers can now spare the land to run pigs outdoors. However, young breeding gilts and pregnant ('in-pigs') sows are better kept outdoors as long as possible and we are particularly fortunate locally in having a farm where one hundred or so run loose during the warm months of the year. The land is divided into about a dozen pastures, which are rested in turn when everything green except unpalatable weeds has disappeared. The pigs are carefully 'wormed' before going on to the pasture (and their piglets will not follow them), because once land becomes 'pig-sick', or worm infested, it has to be ploughed and rested before it can be used again.

Through these pig pastures the stream meanders so conveniently that water is effectively laid on everywhere and it is sufficiently shallow for the pigs to wade in with a certain ponderous dignity wherever they want. Pigs are naked animals, whose bristles are no substitute for a good coat, and they need shelter. This is provided in the shape of simple huts with curved corrugated iron roofs on wooden floors covered with straw bedding, placed so as to face the sun. And when pigs are not actively digging, they devote the same gusto to sleeping. The wild animal will cat nap, with one eye half open, even in sleep alert for danger. A sow, with her belly full, asleep in the sun, is distinguishable in the eye of the observer from a sow who is dead only by a subtle perception that she is enjoying herself. I remember one of these huts, one warm afternoon, in which several sows were asleep across the door in the sun, when one of their companions at the back rose to her feet, clearly feeling the call of nature. She saw the problem at once, and equally clearly deplored the necessity which faced her. With evident regret, she placed her sharp hooves firmly on the nearest recumbent body, heaved up her twelve or fourteen stone and walked over her companion into the open. The only reaction was a momentary twitching of the recumbent sow's large pig ears.

Pigs are not fussy feeders and amongst their main feeds are

barley or oat meal, fish or soya meal, skim milk, bread, potatoes, roots and concentrates. It is the soya, fish or meat meal which provides the essential protein. In this country barley is the main feed and in the United States of course 'corn', which we call maize. But pigs need considerable amounts of vitamins A and D and at different ages carefully regulated amounts of the minerals calcium, phosphorus, salt, iron, copper, zinc and manganese sulphate. Minerals can be bought in balanced mixes and the proportions in a diet for porkers might be for instance, 6 cwt barley, 2½ cwt 'weatings', ¾ cwt each of fish and soya meal, with 14 lb of 'pig minerals', together with 2 million units vitamin A and half a million units vitamin D. (*All About Pigs*, Farming Press).

Although pigs will eat most of the things put in front of their noses, the farmer who is producing pig meat commerically is necessarily most concerned with the best possible conversion of feed into flesh and will probably buy in ready-made feeds if he is operating on any scale. 'Pig starter meal' will contain eighteen to twenty per cent of protein for piglets from two weeks to just after weaning at eight or nine weeks, when they should weigh about 40 lb, and at three weeks the piglets are probably given some food in a 'creep', which their mother cannot reach. (The same method is applied with lambs and their ewes. A creep neatly describes an area which the baby can get at but not the full grown animal.) Grower's meal and sow meal meet the needs of the growing animal up to about eighteen weeks, or of the sow before 'farrowing' (giving birth) and while suckling. The fattening meal used from a weight of 120 lb to bacon weight of about 200 lb need only contain 12 to 13 per cent of protein.

It is the protein of the feed, as against potatoes, stale bread, roots, cabbage or swill, which is expensive, although essential to young animals and in late pregnancy or during suckling. In the first month of pregnancy, the total feed will be 6 to 7 lb a day, dropping to 4 or 5 lb in the next two months, rising again to 7 or 8 lb in the last month. As the litter grows up, the sow must be allowed extra for each piglet she is feeding, up to a maximum of about 14 lb a day, or three times her mid-pregnancy feed. The aim is to raise the litter, while maintaining her condition but not encouraging her to put on flesh and here, as with the cowman and the shepherd, the pigman's judgement and experience are vital to success. The same theme runs through the whole of this book. The margin between profit and loss is minimal. The warning reads: 'Don't let your sows eat your profits'.

The sow and her boar are likely to be one of the popular breeds, Large White, Landrace or Welsh. The Meat and Livestock Commission is responsible for Elite and Accredited Herds and commercial members of the National Pig Breeders Association are energetic in advertising their 'top boars' for breeding, the scores achieved by their boars in MLC performance tests, their membership of the Ministry's Pig Health Scheme, the success of their sows and gilts in averaging for instance, over eleven pigs per litter and rearing ten of these, or their contribution to improving the 'food conversion rate and lean meat content'. All this is a reminder that we are breeding for a whole gamut of qualities: good mothering, rapid growth rate, metabolic efficiency in converting food into flesh and, at the end, the sort of carcass the butcher likes.

A sow comes on heat again a few days after weaning her previous litter and the cycle of mating readiness is repeated every three weeks. Gestation is 112 days and there is an interval of five days or so between weaning and the next mating. Thus the sow produces two litters a year, but even so, commercial needs have pressed some pig breeders towards weaning soon after birth, or when the piglets are a fortnight to three weeks old, instead of leaving them with their mother for the traditional fifty-six days. If the standard of management is so high that this can be done successfully, then the sow will be started on her next litter so much earlier. And this explains the higher rate of feeding in the first month of pregnancy, even with normal weaning at fifty-six days, for it helps the sow to recuperate after suckling the previous litter. If, after eight good litters, she farrows two successive poor ones, then her breeding life is probably over. It will have started when, as a gilt, she first went to the boar at eight months old.

'Food conversion' depends on the breed, but this is what the cost-conscious farmer watches. A table can readily be compiled showing the weight of pig from say, 40 to 200 lb, its daily poundage of food at each weight, the daily gain in weight expected of it, and as a result the conversion factor, ranging from 4.1 when it is a 'baconer' at 200 lb (or higher if it goes on to become a 'heavy hog' at about 260 lb) to 2.3 when it is a little forty pounder, converting 2 lb of food daily into between ¾ and 1 lb of flesh. The greatest skill lies in breeding, and if the pig farmer has produced successful litters, he will make more if he can go on to fatten them as porkers, or beyond that as baconers, instead of selling them as little weaners to a specialist fattening enterprise. But he must remember that what is expected of him are highly standardised carcasses, just as the vegetable grower is expected to produce crops of uniform size and quality – and in pigs 'a white skin is preferred'.

The pig was the village cottager's animal. As Flora Thompson describes it in *Lark Rise*, 'During its lifetime the pig was an important member of the family, and its health and condition were regularly reported in letters to children away from home, together with news of their brothers and sisters. Men callers on Sunday afternoons came, not to see the family, but the pig, and would lounge with its owner against the pigsty door for an hour, scratching piggy's back and praising his points or turning up their noses in criticism.' Children and all contributed to his feeding and when the great day came for his slaughter, there was a feast in cottages which seldom saw fresh meat. Then, after the 'pig feast', his bacon saw them through the winter – hence, no doubt, the phrase 'saved our bacon'. That background from late Victorian times was still in the minds of those who after the Second World War called for 'a few pigs on every farm'. But today there is no profit in a dozen pigs, and you cannot keep a pig in the garden of a council house even in the country, although a farmer may quite well run a herd of fifty sows or so, fattening the 1,000 piglets they will produce in a year himself, and this is a convenient one man job.

Specialist pig fatteners now work on a large scale, with a heavy capital investment, buying in weaners from specialist breeders or from smaller farmers who for one reason or another decide against carrying them up to slaughter weight on their own farms. Now you may find 1000 'fatteners' in a single enclosed house and a large piggery may run to two or three such houses. It is a far cry from the Victorians and this is where you encounter factory farming, with automatic or semi-automatic feeding, controlled ventilation, the problems of water supply and dung disposal on a great scale, the need for extreme care to avoid contagious disease amongst animals so concentrated in a small space – and output on a scale to meet the rapacity of our demand as consumers. We have lingered with the pigs mainly as an example of intensive 'indoor farming'. It also applies to cattle. Unwanted calves are bought from dairy herds at about a week old and drink warmed milk, or a substitute, from a mechanical udder until they are about three months old. One man can care for several hundreds on this system. They may be kept indoors and fattened until they can be sold for beef, although this has become costly with the great increase in the price of the barley which was the main feed. It also applies above all to poultry.

In this factory farming contagious diseases can of course be disastrous. They may involve losing not only the entire population of animals in the house at the time, but also the cost of completely disinfecting it and possibly allowing a period to elapse before it can safely be used again. Then you need piped water in great quantities. To go back to pigs, each breeding sow with two litters a year needs about 1000 gallons. Above all there is the problem of slurry. This arises simply when large numbers of animals are kept indoors on bare floors, 'hundreds of acres of concrete', often with slats in the floor, through which their dung and urine pass into underground conduits and tanks. When dung and urine are dropped on to straw they mellow into farmyard manure. Without straw there is slurry, and although it contains the three basic plant nutrients, they are not easily used in this form. Also pig effluent is very unpleasant and about three times as strong as human sewage. And the quantities are very large: 750 baconers in ten months would fill a slurry lagoon 100 feet by 75 feet to a depth of 6 feet. Moreover, bacterial action can in no way be relied upon to render it innocuous.

The general farmer can use his pig dung mixed with his straw as manure; in fact some potato growers value pigs for their manure. Also there are experiments in turning dried and ground poultry droppings into a food which cattle and sheep can use for its nitrogen. But the specialist with only pigs kept in closed houses without straw and no other use for their droppings is reduced to spreading it over such land as he has from great tankers handling 350 to 1000 gallons at a time.

Factory farm housing is easily recognised by its varying types of ventilator and the houses are generally windowless. They are recognised too by the upstanding hoppers, from which ready mixed feeds drop down by gravity into distributors. Pigs are not finicky feeders, and it may be convenient to distribute their food

17 *A Suffolk type pig-fattening house. The bales of straw are for insulation. Bright and intelligent faces emphasise the danger of boredom*

in liquid form by pipes supplying their troughs. But pigs, like old folk, must be protected from cold and draughts, and their housing needs to be insulated to prevent condensation. The slogan is 'snug but fresh' and low ceilings keep the warmth down where the pig wants it. Pigs favour very smoothly finished concrete floors and kennels with lids about $3\frac{1}{2}$ feet from the floor, so that their body heat remains around them and is not dissipated up to the roof. Generally the kennel lids are further insulated by heaping straw on them. One thing you can be sure of with healthy animals is that they will not soil their bedding, but will dung in the area provided, whether on straw for manure, or on bare concrete for the tractor with its scraper, or through slatted floors into the slurry pit.

Another common feature is that just as the bull (if there is one) wants to see what is going on, so it is only thoughtful to locate the boar's pen between yards occupied by sows. But in this indoor farming there are two alternative principles. Poultry may be held in individual cages, or laying hens, at least, may live in large groups wandering at will in big houses on deep litter. Calves may be reared in individual stalls, but beef stores being fattened indoors will be grouped in communal pens. Cows may have communal, strawed bedding areas or individual stalls in which to rest and ruminate. In either case they will have passages in which they can wander at their fancy and will feed communally from long troughs, from communal, centrally placed hay mangers or from silage heaps. Similarly, pigs may share common yards, but enjoy private bedding areas. Ewes wintering indoors will be wearing their winter fleeces. They will be grouped in pens, in much more flimsy housing, roofed, but with walls which do little more than keep out the wind. Whenever birds or beasts are confined in groups, then they should be of an age and have grown up together. A strange child introduced into a class of established schoolfellows will be under no more strain than a strange cow in a well adjusted dairy herd and there will be fighting between mixed groups of pigs or poultry.

The physical lay-out of housing for cattle, pigs and poultry will often be the same in principle, with stalls, kennels or cages either side of a central passage flanked by their feeding troughs or containers (unless the pigs feed off the floor) and access at the other end of the animal for the tractor to scrape up the dung, unless there are either slatted floors or deep straw in which the dung accumulates. With caged birds, the droppings can be carried on a conveyor belt and the hen's eggs can be made to roll down a gentle incline to be collected from trays lining either side of the central passage.

The basic question is whether there shall be communal feeding, as in a ship's or regimental mess, or individual feeding. With hens the choice is again between deep litter houses and individual cages. Unfortunately, if sows or gilts are fed together, particularly on open pastures, it is the quick eater who will get so much that the slow or shy one goes short of her due. Sadly too, the specialist journal *Pig Farming* reports the growing violence of our time: 'Savaging among sows seems to be on the increase' and can be fatal. Most fighting takes place over food and it seems that with pigs, as with poultry, there is fighting to establish priority, a social hierarchy or 'pecking order'. Another heading from the journal is 'Feeding without Fear' and the implication is towards individual feeding, which implies individual kennels and only common exercise and dunging yards. With indoor farming there is also the menace of boredom, at least with such naturally active animals as pigs, and where hens may go for each other's tail feathers, pigs are liable to periods of tail-biting, or teeth grinding. Fortunately their attention can be diverted by giving them a shovelful of small coal to chew, which seems to improve rather than upset their digestion, or they may be given a few turnips or even a paper bag or two, or a few plastic footballs to play with. (This suggestion, again from *Pig Farming*, seems to bear out the old saying, probably dating back to the days of Flora Thompson's cottagers' pigs, that 'dogs look up to you, cats look down on you, but pigs is equal'.) Nevertheless, there seems no way of treating woodwork so as to prevent pigs gnawing at it wherever it is exposed.

A yarded sow will need about twelve square feet for sleeping and eighteen square feet for dung and exercise, but the farrowing sow presents a special problem, probably because she has been bred into such a specialised piglet producing animal, just as we have selected hens as egg layers to the point at which they are so brainless that they are liable to starve in sight of food, if the food is unfamiliar to them. Sows normally suckle lying down and the trouble is how to prevent a creature of, say, eighteen stone flopping down or turning round and so suffocating one of her numerous litter of little piglets. In a farrowing pen or crate, the sow, while her piglets are young, will be confined between rails in a space about six-and-a-half feet by a little under three feet, but the piglets can slip beneath these rails. They can also slip out of the sow's stye altogether, through a creep which is far too low for the sow, just as lambs will slip through creeps on to fresh grass which the ewes cannot reach. In the creep, the piglets will learn to enjoy their piglet 'creep feed' and they are attracted into the creep by giving it heating lamps which make it warmer than the sow's stye. It is when they are not warm enough that they will nestle up to mother instead and may be accidentally crushed when she turns.

Generally, the piglets get an iron injection when three days old and the unfortunate males may be castrated at the same time, or later. Normally it is not necessary for the pigman, unlike the cowman or the shepherd, to be present when the sow farrows, unless she is known to be careless and apt to lie on her young. But he should be at hand when a gilt is having her first litter, because it is a new and alarming experience. She may take fright at the sight of blood and if a frightened gilt shows signs of savaging her young, it is not uncommon for the vet to anaesthetise her. When she comes to, she will mother her litter without further ado.

Enough has been said to show that the dairy or sheep farmer, the beef and pig grower and the poultry farmer earn what profit they make. But to complete the survey of factory farming we must turn from the glorious gluttony of the pig to the pallid immobility of our poultry.

The 'broiler chicken' was unknown in 1950. Today the number consumed must be nearly 300 millions in a year. Before 1939 poultry flocks existed on general farms to use up the tail corn and to help improve the grassland. Today, despite numerous advertisements in cottage windows, free range poultry flocks have virtually disappeared from farms. The change however, has not been without its difficulties. 'Most . . . retailers must have . . . brown eggs and this consumer preference shows no sign of weakening. A few years ago most of the brown-egg laying breeds tended to be larger type birds eating more food than their white-egg counterparts. They also tended to lay fewer eggs in the laying year. However, in response to demand, poultry breeders have improved the performance of many brown-egg strains and some of the best are now comparable with the white-egg laying breeds.' (Fream's *Elements of Agriculture*.)

Pullets lay for about a year and then go into a moult for six to

18 *Breeding hens. The droppings fall through the wire floor*

ten weeks, during which they are unproductive. If they survive into their second year, their output will fall by about one third. The egg producer now often buys pullets from specialist breeders at point-of-lay and these will generally be already vaccinated against fowl pest and infectious bronchitis. Moreover, it is possible to have the performance of specimen birds impartially tested. If day old hen chicks are bought it will be four to six months before they start laying and those which grow up during the period when the days are lengthening mature more quickly than those reared as the days draw in, because light stimulates the glands which promote sexual maturity. A dim light helps to cut out feather-pulling and cannibalism amongst growing pullets. As they are easily upset, it is best to rear them in the same home from the first day until they are ready to lay.

When they are laying, either in batteries or on deep litter, they may be fed mash or pellets, but the pellets help to reduce boredom. Amongst these close-packed nervous birds, the mere dropping of a sack can create panic, so that a great deal depends on the quiet and calm with which the poultryman carries out his tasks, and you are unlikely to be shown over an enterprise of this sort. The amount of light makes little difference to the birds, provided the man himself can see what he is doing, but the *duration* of light is crucial. The longer the day, the greater the number of eggs. Research has also shown that hens react best to an increasing length of day. The day is kept relatively short until the birds have reached maximum output when they are about thirty weeks old. Then their artificial day will be increased by fifteen minutes a week until it reaches a maximum of about twenty-two hours. This interference with the earth's rotation delays maturity by two or three weeks, but produces slightly larger and more eggs throughout the laying year. It follows that the houses are windowless and artificially ventilated.

'Broilers' are sold when eight to ten weeks old, weighing 3 to 5½ lb. More than four-fifths of the producer's outlay goes on purchasing the chick and its food. The price of the chick and the selling price of the broiler are both things he cannot much influence, so he concentrates on reducing food costs if he can. This turns, as usual, on the food conversion rate and 2·3 lb of food to 1 lb of meat is now regularly achieved. Artificial light is now also normal, and a dim red light seems to quieten the massed birds, but it is essential that they are never left to panic in sudden darkness.

Specialised poultry enterprises of this sort, unbalanced by any other type of farming, invite the comment that the farmer is risking all his eggs in one basket. The capital involved is also very great. A 20,000 bird house is now about the minimum and the egg trade is dominated by very large units of about 100,000 birds. The farmer with 3000 to 5000 birds on deep litter in an old barn, stable or loft will make the enterprise pay only if it is a part-time occupation and his eggs are sold at the farm gate direct to the consumer or through his own milk round.

Turkeys are also mass produced in the same way as broiler chickens. The annual output is around 15 millions and the trade is anxious to popularise turkeys throughout the year. This involves trying to produce mini-turkeys, for birds of 16 to 24 lb are inevitably associated with Christmas fare. The poultry which used in general to be a luxury is now one of the cheapest meats.

All this has brought us a long way from the open air pigs whose uninhibited porcine ways occupied the beginning of this chapter. No farm animals lead natural lives. Yet it is difficult to deny that the life of pigs in intensive fattening houses and particularly of caged birds in broiler or egg-producing houses is different in kind from anything we have known before. Eggs can be produced by hens on deep litter and even the layman who only observes the work on the land, or the gardener, feels in his bones that it is right to return to the land as manure the fertility which the land provided. Hens on deep straw and pigs in less intensive systems do this, and unless slurry can be turned into something positively useful, it remains suspect.

But this partial alternative to battery hens for eggs does not help the broilers, who in their brief lives have made so little use of their bodies that when they are cooked the bones almost fall out of them. And it is poultry and pig meat which, roughly speaking, has replaced the beef which we used to import. To demand the abolition of factory farming in a hungry world is pointless if we cannot get enough meat and eggs without it. Even the vegetarian offers no alternative. At present it would hardly be possible to produce enough high-protein vegetables, eggs and cheese to replace what we get from these factory farms; and any solution which is only open to the few and will not work for society at large is no solution, but only hypocritical. We can only insist that the conditions in factory farming of any sort do not cause positive suffering to the creatures involved. This our present regulations claim to do. If there is evidence that they are inadequate, then we must make the regulations more exacting and pay more for our food, or eat less.

In the meantime, factory farming will spread and with it the growing of vegetables under cover in a controlled environment. It may powerfully help to change our unnatural thinking of the past century or so; and under one hundred years is not a long period in history. During this period we have regarded ourselves as dominant exporters of manufactured goods and of sophisticated services, importing in exchange cheap food and raw materials from less developed countries. We no longer dominate in industry and there is no longer any cheap food to import. We are moving and must move towards feeding ourselves again; and we should soon be thinking of ourselves as a food exporting nation.

7

THE UNNOTICED REVOLUTION

By the late Georgian period, about the end of the eighteenth century, it was becoming the practice to bring the cattle off the fields as the year declined, and to in-winter them in yards and later in cattle houses or covered yards. There was money then in the pockets of landowners and farmers, if not of their labourers. And there was still money in farming after the repeal in 1846 of the Corn Laws, which had given protection against imported grain, right up to the collapse of British farming in the last decade or so of the nineteenth century, when the grain of the North American prairies, the tinned and later refrigerated meat of Argentina, the meat, butter and cheese of Australia, New Zealand and Denmark began to come flooding in. The original buildings which form the farmstead of the typical lowland farm often date from that period of about one hundred years. They were set up and indeed carefully designed for the horses and cows, the bulls and bullocks, the pigs, the hens scratching about the yard, the ploughing, threshing, storage and dairying of that period.

Sometimes, quite often in East Anglia, you will find an older barn, going back to Tudor times or earlier and possibly still thatched. Inside, this is like a medieval church, or perhaps even more like one of Henry VIII's great wooden ships turned upside down, its massive oak beams and struts lost in the dim light far above your head. But, broadly speaking, it is the buildings of the high Victorian period which end an epoch. After that, until recently, there was no money in British agriculture. The home farmer who in the 1870s had fed three-quarters of his countrymen, by the early 1900s supplied only one quarter of their food. The Great War of 1914–18 threw us back on our own resources, to supplement what could be brought in past the German submarines, but after the war was over farming collapsed again. Fields which had been sown to corn during the emergency tumbled down to grass again. This was a period of 'dog and stick' farming, when it often did not pay to plough, and a man grazed his stock on what rough grass they could find and an acre of land could be had for £20. Sometimes those who got off best were Scots, who had farmed thin upland soils in the north and were able to do a little more than scratch a living from the good lands for sale at bargain prices in the south.

Then the Second World War, and after that both political parties were determined that Britain should not again be as dependent as she had been on imported foods and on the self-sacrifice of the Merchant Service and the Royal Navy in bringing them in across hostile seas. Agriculture certainly never attained the same position as the armaments industry, where the taxpayer's money went in 'open-ended' commitment to projects whose cost increased in geometric proportion as they progressed. Prestige and the magic of national defence make it seem only natural to governments that as soon as one 'generation' of war machines has been completed, work should start on another 'generation', infinitely more costly, to supersede the first. The vagaries of human psychology react less generously to food production, but British farming has been supported, albeit less generously since the mid 1950s, and has responded in a revolutionary way. The bumper harvest of 1972 marked a high point in our long agricultural history.

There have been two revolutions. The first results from the tractor, the bringing of electricity to the farms and from then on an ever increasing complexity and power of machines, electrical or propelled by diesel engines, with modern hydraulics and ingenuity. The caterpillar tractor derives of course from the tank of the First World War. The second revolution, of quite recent years, emerges from the laboratories and experimental farms. This divides into chemistry and biology. Advances in chemistry have led to new control over weeds, plant diseases, soil pests, and, with veterinary science, over animal pests and diseases. Agricultural botany has contributed to weed control and to all the improvements in cereals, grasses and other crops. Genetics and the latest 'progeny testing' have improved livestock breeds. This second and recent revolution has made farming the sophisticated and precise operation which this book has tried to describe.

The weeds have almost disappeared from the crops; the yields have risen enormously, as has our degree of self-sufficiency. The tractor, the combine, the sugar beet and potato harvesters – and the disappearance of the old 'agricultural labourer' – have necessarily cut out hedges and thrown several small fields into one large one. The plough has climbed the hillsides and often turned the chalk slopes of the Chilterns and the South Downs, the slopes of the Pennines and the Yorkshire Moors, the uplands of the Border country, Wales and Scotland into arable fields and leys, which before offered only sparse grazing. But rain, mud and dust persist. In an age when every mechanic wears both the white coat and the air of a clinician,

19 *The tractor now does the heavy work – stacking straw to remain in the field*

the farmer and his men still go about in gum-boots and a worn old jacket. Perhaps it is this which has helped to disguise from an urban population that his revolution in its way has been as remarkable as that which has produced the transistor radio and that, for the most part, farmers are aware of it.

I take then as typical a farmstead designed in 1849 for a mixed farm of 300 (now 800 acres) and try to trace how new needs, new knowledge, new machines, the departure of both horses and steam engines have altered these buildings, added new ones to them, in part left them deserted; and have changed the farm they serve. Let me describe the buildings, which were most carefully thought out and show many detailed improvements on the farmsteads of fifty years earlier, although in essentials they remained unchanged. The actual farmhouse, the farmer's dwelling, lies away to the south-west; that is, in this country generally up-wind of his livestock and their manure. Their yards follow the by then well-established pattern of being open towards the south. In the centre of the row of buildings along the northern side is the engine house, with its tall chimney; next to it the threshing machine which the steam engine works, and above them the granary. Conveniently adjacent is the barn for the threshed straw, to its right the covered house for the fattening steers who will use the straw and to their right the feed store. Round the corner beyond this is the part of the building allotted to feed preparation, next to it the cow house and beyond that the calf house. In front of the fattening house and the cow house are the walled cattle yards and between them the manure yard.

Go back to the engine house and the straw barn in the centre of the northern wing and moving now to the left there is the storehouse for turnips and other root feedstuffs. On its left the hay barn (all of these were at that time solid brick-walled buildings). Turn the corner and you have the blacksmith's shop, followed by the long stables for the farm horses and the riding horses, with their yards in front of them and the house for the farmer's gig beyond them. Here you have, logically arranged, an E shaped run of buildings, open to the south, the west wing

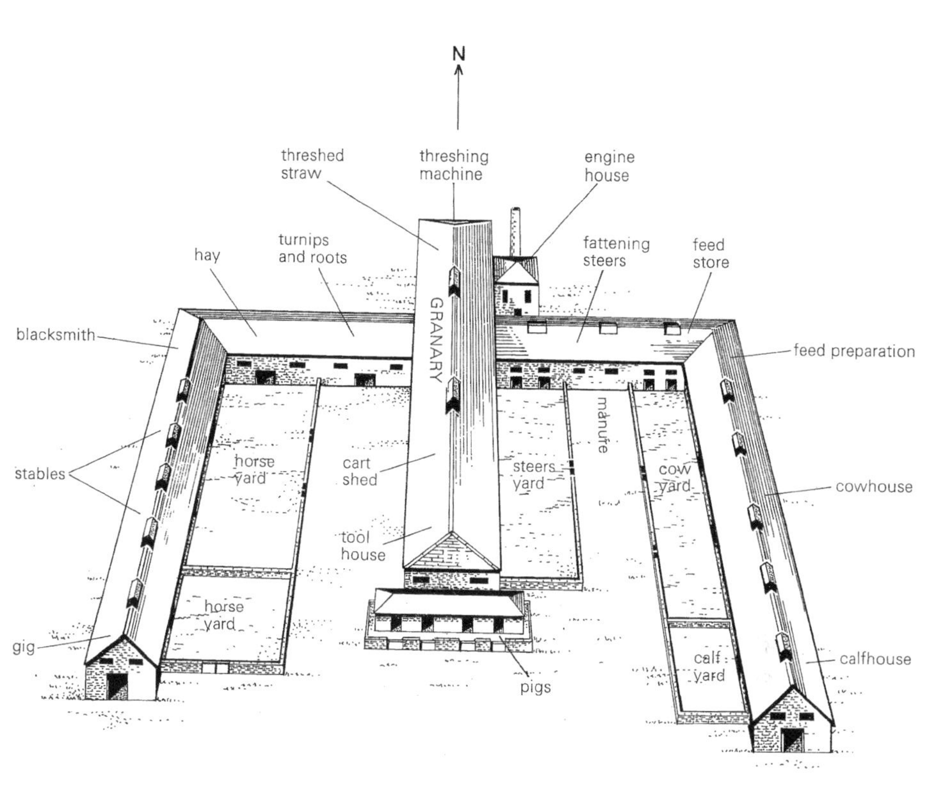

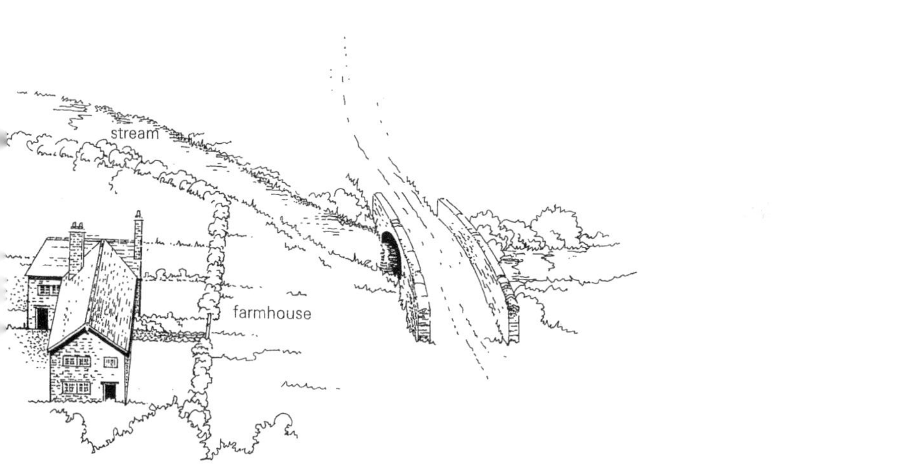

Adapted from *A History of Farm Buildings in England and Wales* by Nigel Harvey (David & Charles, 1970)

housing horses, the east wing cows and their calves. The centre bar of the E, dividing the horse yards from the cattle yards, consists of the pig styes, a tool house and a cart shed. The hens are what we have learned to call 'free range' and the dairy is part of the farmhouse. What strikes us today at once is on the one hand the solidity of the whole establishment, which accounts for much of it still being there, and on the other the underlying assumption that the established order of things would continue unchanged through the generations.

The first change came inside the pipe bringing mains water to the farmhouse and on to the troughs in the cattle sheds and horse stables. Pumping water up from the well stopped; the old horse pond became overgrown and the muddy track down to the little stream meandering between the farmhouse and the steading was hardly used. Not only did the stock have fresh water readily available, but cleaning out the cow and bullock sheds became at once less laborious. Behind the water came the onward stride of the poles carrying electricity across the countryside. Between 1939 and 1960 the percentage of farms with mains electricity rose from a mere eleven to eighty per cent and farmers were not slow to increase their consumption of current. To us this means the immense convenience of light at the flick of a switch, but to the farmer – the current reached our farm shortly after the war in 1948 – it meant power as well. The old machine for crushing oats and barley for cattle feed, with its heavy cast iron wheel turned by hand, was pushed aside, although it can still be seen alongside the new electric motor which now does the job. The laborious business of chopping up roots for feeding also passed to a separate electric motor. Between 1939 and 1960 the number of electric motors on British farms shot up from 11,000 to 253,000.

But sensational as this seems – and it shows that farmers were no laggards in change – the greatest changes came on the wheels of the tractor and the combine. In 1939 there were 725,000 working horses on British farms. Now there are virtually none. About 50,000 tractors were already working by the outbreak of war. In 1970 over 400,000 much more powerful ones were at work. By 1938 about 100 farmers in the British Isles were already experimenting gingerly with combine harvesters. In 1970 66,000 much larger ones were reaping the entire harvest. But combines mean grain driers and they multiplied threefold in the decade 1960 to 1970. Mechanical potato harvesters numbered 3000 in 1961, but only six years later had increased to 26,000.

This was revolution on our farmstead. The tractor heaves up manure, which formerly was loaded by hand. The tractor scoops up grain, lifts bags of fertiliser, shifts bales of straw and hay, loads potatoes and an elevator loads sugar beet. Cutting hay or straw from stacks by hand with a knife is a thing of the past, now that the new machine delivers them in bales. You have only to notice the size of the farmworker's chest and shoulders to realise that he still uses his muscles, but the routine heaving of loads by the hundredweight is finished. If a man is humping heavy weights by hand on our farm now it is a sign of stress, of the organisation to some extent temporarily out of gear.

But the tractor and the combine have changed much more than this. There is no longer any horse manure. The long row of stables became silent and empty, except for rubbish, although twenty years later there were still men on the farm who could stand in front of the loose boxes and tell you the name of the horse which was stabled in each, when they first started work. Even the space for the farmer's gig is empty, because he houses his car in a new and not very beautiful garage run up conveniently alongside the house. Equally drastic was the effect of the combine, because it reduced the steam engine and the threshing machine to the status of museum pieces. The rickyard is empty, but what the combine does need is a grain drier. In many old barns you will see a sort of dormer protruding where the drier has been installed. In our case the steam engine and the threshing machine went for scrap. The good Victorian walls on the north side of their housing were knocked out; the roofs supported on steel girders and grain silos built in beneath them – a row of three gleaming metal cylinders under the shadow of the old eaves. These are vertical flow ventilated bins, in which the grain, lying anything from five to fifteen feet deep, is dried by slightly warmed air blown upwards through the porous floor from bottom to top. Along with this goes the mechanical handling of the grain, and a mechanical elevator, again with its own electric motor, to stack the bales of straw and hay. The north side of our range of buildings is now a fine architectural example of new wine in old bottles.

But the idea of walled-in storage for hay and straw was always rather peculiar and the new open sided Dutch barn towers up to the east, at right angles to the cowshed. It is almost as long as the farmhouse itself, and a good deal taller: an arc of tar-coated, corrugated iron roofing, held aloft on great steel girders and with brick walls half way up the two ends. Parallel

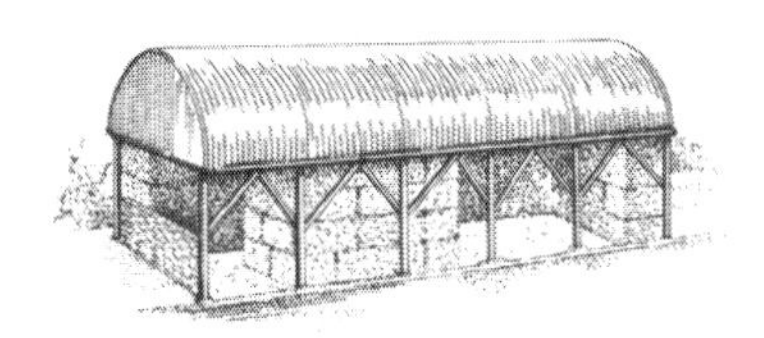

The great Dutch barn

to it is the new tractor shed, again a corrugated iron roof with light brick walls on three sides and an open front. The farm now houses three combines, each in its own Nissen hut type of shelter, a curved roof resting either side on five feet of brick, with a brick back and an open front. Simple and cheap, but impressive alongside the tractor shed, simply because there are such enormous machines. The old buildings were all designed for the movement of men and beasts; the new buildings allow for the easy manoeuvre of huge machines.

Clearly their effect is not restricted to the farmstead. It is far greater on the fields. Following a pair of willing horses a man could plough one acre four to six inches deep in six hours. Today a large tractor will plough an acre almost twice as deep in half an hour, although it must be admitted that some of the most powerful of the modern machines can seldom be used at the full rating given by their manufacturers, simply because men cannot stand the noise and vibration of the engine and transmission at full throttle. These machines have come to us from the open prairie and it was to be expected that they would disrupt our small enclosed fields. Hedges have had to go, and not only for mechanical reasons, but also for the human reason that men are no longer available for hedging and ditching. Open ditches are piped and filled in; space is gained (some compensation for thousands of good acres lost to buildings and to sometimes megalomaniac motorways and roundabouts); labour is saved, and incidentally crops which were formerly overshadowed by hedges are now open to the sun. The problem is that what may (since the eighteenth century enclosures) have been five fields with five different names in now only one. Where you had Lion Field, The Down, Long Acre, Highlands and Parsons Fall, somebody had to decide that the new single field should be known as Lion (after the Black Lion pub at its

20 *A massive, four-wheel drive, articulated tractor at work with a seven furrow, non-reversible plough (which is off-set, so that all four tractor wheels are 'on the land', instead of having the right-hand wheels in the furrow). This is equipment for large fields and large farms*

foot). The open ditches that remain, after their winter cleaning, look as if some giant hand had been at work on them, but again it is only the tractor fitted with a mechanical digger. The loss of shelter to hedgerow birds and to the small animals of the countryside is also a loss to us. There is some compensation in the rows of thin new saplings which have recently been energetically planted along the road side. At intervals are concrete 'aprons' for the dumping and handling of manure, sugar beet and other crops, another characteristic sign of mechanised agriculture.

Some of the new machines are delicate. The drills for corn and other seeds, the big sprayer and above all the delicate specialised drill which enables the farmer to drill regularly spaced seeds 'to a stand' all need to be properly housed. These too have brought in a minor revolution, because the precise placing of seeds and fertilisers, together with pre-emergence weedkillers, cut out the need both for thinning the crop and hoeing down the weeds. Young weed seedlings and germinating weeds are killed by a chemical to which the crop itself is immune, or the crop is protected by the depth at which it is sown, and the weedkiller's residual effect keeps the soil clean for some weeks. Or the weeds are killed by some appropriate chemical sprayed on their leaves and again it must be one which does not harm the crop. The old row of stables, opened up and cleared of internal divisions, now houses these instruments of very modern technology, and the ploughs and harrows share the same shelter.

The chemist and the National Institute of Agricultural Botany in Cambridge have wrought great changes in grasses and in cereals. Just one example, beyond what has already been discussed, of the application of science to grass. You want the optimum quality and digestibility over as long a period as possible. The Institute issues a Prediction Chart for silage cutting at '63D value', which means that 100 lb of dry feed contain 63 lb of digestible, energy producing, organic matter. This is accepted as the right moment at which to cut and different grasses are specified which reach this value at different times, (e.g. mid-May for Ryegrass S24 and mid June for S23, which is why both S24 and S23 are included in the mixture mentioned in chapter 3). Again, cereal seeds can be protected against seed-borne diseases like Leaf Stripe by a dressing with organo-mercurial seed powders, and if it is known that the field has a high population of wireworms (most ploughed-up grassland has), then the seed can be protected in advance with a combined dressing against insects and funguses.

But it is breeding improved varieties that accounts for a large part of our increased yields and an acre of wheat will now produce twice as much grain as it did thirty years ago. In the last generation the French and the Belgians have successfully bred for greater strength in the straw. This at once allows the use of more nitrogen fertiliser, which results in a greater quantity of grain, but without the risk of the plant 'lodging', or keeling over. The scientists have discovered that the Yellow Rust disease consists of several biological strains and have bred cereals resistant to all, as well as to Loose Smut, mildew, shattering of the ear at harvest or sprouting of the seed in the ear. Both maize and lucerne (alfalfa) are now increasingly grown for winter feed, although maize does not fully ripen in our climate. Lucerne is especially useful in the drier counties, because its deep roots resist drought. Unfortunately it develops very slowly, but then its seed can be innoculated with bacteria which develop nitrogen-fixation nodules in the seedlings' roots. This establishes the plant much more quickly and once established it will give a good forage crop for a succession of years. One further example of the NIAB's work: it will specify the best variety of sugar beet to grow in each part of the country.

When our farmstead was built, in early Victorian times, labour was cheap and the country grew over two million acres of turnips, swedes and mangels as winter stock feeds. This accounts for the special turnip storehouse, but as long ago as 1958 the acreage of these root crops had dropped to only one sixth of what it had been and our turnip house has long been empty. The pig styes too have been knocked down. Farmers no longer keep the odd half dozen pigs: this is a specialised enterprise, to be carried out on a large scale or not at all. A thoroughly modern fattening house for 250 pigs, fully enclosed and with air conditioning has been built elsewhere on the farm. This in itself is an indication of decisive change in very recent years.

Where the old pig styes and cart shed once stood, there is now however, a glass house and its purpose is to expose the seed potatoes in the spring to the maximum daylight and to fluorescent lighting at night, so that they may be well 'chitted' or sprouted in their trays before they are sown. All this trouble is only devoted to the best seed available and the agricultural departments of the United Kingdom and of Eire have a common system of inspecting and certifying seed potatoes. The seller must quote his number and the quality letter he is permitted to use: H for healthy commercial seed; A for the first quality of commercial seed potatoes, and SS for the very highest grade, mainly for the use of farmers who aspire themselves to grow seed potatoes – plus in each case a letter which tells the buyer which county his seed comes from. This is important, because potato virus is spread by aphids. These pests will not fly if there is a wind blowing at more than 4 m.p.h.; if the temperature is below 65°F and if the humidity is above 75 per cent. These conditions happily apply to the Scottish and Irish districts where the best seed potatoes are grown, as well as to many of the hilly parts of England and Wales.

It is worth at this point discussing what one writer confidently calls 'Britain's Lead in Farm Research'. Successive

British governments have spent proportionately more on agricultural research than any other European country and it is to our research stations that we largely owe our undoubted technical lead. There are some forty of these, under the Agricultural Research Council, linked of course with research in universities, and tending to specialise in different areas. These vary from that of the Grassland Research Institute at Hurley (there is also a British Grassland Society) and the Redesdale Institute in Northumberland, concerned to improve the economics of hill sheep farming (and there is also a Hill Farming Research Organisation), to the still more specialised work of the Wye Hop Research Institute and that of Martyr Worthy in Hampshire, which is concentrating on the several problems which have emerged from the present economic drive towards running very large dairy herds. The Reading National Institute for Research in Dairying has made a very practical contribution in a method of diagnosing not the acute mastitis, which is obvious to the cowman, but the often concealed sub-acute condition which, because it escapes notice, causes a great deal of loss.

Some pests, such as the Colorado Beetle affecting potatoes, have to be notified to the Ministry, which will supervise spraying and soil fumigation and regularly inspects suspected areas. Some insects are now destroyed by 'systemic' insecticides. These enter the plants' sap and kill off any insect which unwisely sucks the sap stream. Up and down the country, the Experimental Husbandry Farms evaluate the results of this research on a practical scale – and many motorists touring the countryside must be baffled by the initials EHF, followed by a name. The Agricultural Council and the Royal Agricultural Society (dating back to the nineteenth century), together with the Ministry's advisory service are responsible for publicising the new ideas.

Modern soil testing makes it possible to make good quite minute but important deficiencies in trace elements. For instance, swedes fertilised with a manure containing the trace element boron are less prone to the 'heartrot' from which they suffer in soil containing a high level of phosphate. More important, research shows that lack of trace elements such as copper, iron, manganese, cobalt and zinc, particularly in some districts, sharply increases the likelihood of sheep falling prey to parasites, and when these deficiencies in their feed are made good their health and vitality pick up at once. We have learned too how to cope with the 'grass staggers' or hypomagnesaemia. This is due to a sharp fall in the level of magnesium in the blood. It affects cattle and sheep and is the more distressing because 'sudden death may be the first sign of the disorder', although there is usually some warning in restlessness and nervous twitching, leading to convulsive fits. It is tackled by including magnesium in the feed or by top-dressing pastures with limestone containing magnesium. Unfortunately there is as yet no vaccine effective against all the different strains of foot and mouth disease in cattle, sheep and pigs. This is endemic in many countries abroad, but occurs here only when the infection is inadvertently carried in, and the policy is to slaugher all infected animals. It is still a matter of controversy, but the official view is that this is better than a suppressed degree of permanent infection, such as occurs, for instance, in France under vaccination.

Calves are now dosed against lung-worm before they go out to graze. We can also protect our grazing livestock from the warble fly, called the gadfly, because the tormented herd goes 'gadding' or stampeding across the field, if the unchecked larvae of these insect pests burrow into their skin and eventually raise great lumps on their backs. Rothamsted, the doyen of experimental stations, is at present working on the hope of extracting protein directly from the plant leaf, as a means of making good the world shortage of protein foodstuffs. A Unit of Nitrogen Fixation is also attempting to enable plants, with the aid of bacteria, to synthesise naturally what fertiliser manufacturers now produce with great quantities of ammonia, nitrates and power. Nearly thirty years ago, it was revolutionary work at Rothamsted too which proved that yields of wheat, barley and mangels were just about the same whether the field had been ploughed or not, provided the seed bed was clear of weeds. Ploughing and harrowing are the traditional ways of cleaning a seed bed of competition by weeds, but chemicals are an alternative, especially since the concoction of Paraquat in this country in the 1960s. With the right soil, corn can be directly drilled into a sprayed stubble or into a grass ley which has been killed off by Paraquat, using a fascinating special drill, whose sharp disc coulters – set in couples at a V angle – cut a neat slit in the compact soil, into which the new seed then slides.

Our own 1849 farm has given up beef cattle and the former fattening yards are now used in winter by heifers destined for the dairy herd, but a book on beef, published in 1972, comments especially on the exciting developments of the last decade. These include quick-fattened 'barley beef', a shortlived fashion because of the rocketing cost of barley; the arrival of the Charolais and other foreign beef breeds; the 'finishing' of Friesian and Friesian-cross 'grass-cereal' beef in eighteen months and, above all, the setting up of performance testing stations.

The Meat and Livestock Commission now incorporates the Pig Industry Development Authority (whose headquarters were always perhaps an incongruous neighbour to those of the Library Association, near the British Museum in London). The

Commision provides beef rearing farmers with precise 'targets' for daily weight increase in these cattle fed on a mixture of grass and cereal, with correct stocking rates at different periods in the growth of the grass, and with 'targets' for suckling cows on lowland, upland or hill farms. The Commission has a Head of Beef Improvement Services; there are lists of top-ranking bulls, boar progency testing stations, a Meat Research Institute and an Animal Breeding Research Organisation.

Britain has been called the world's stud farm and our breeds have in the past been used to improve many of those of today's most important meat exporting countries. Now we are cross-breeding with animals like the Danish Landrace pig or the French Charolais bull which carry in them characteristics derived from native British stock which went abroad in the eighteenth century. The breeder, whether of plants or animals, seeks 'hybrid vigour'. An example is hybrid maize, whose parents were chosen not for their own characteristics, but for the successful way in which their genes intermingled with those of other lines. Similarly, crossing Swaledale and Blackface sheep produces ewes whose lambing performance exceeds that of either parent. The last ten years or so have seen similar changes in the breeding of cattle, pigs and poultry. With poultry of course – improving both meat and eggs – the breeder is closest to the happy situation he occupies in plant breeding. With other animals he has to wait longer, although artificial insemination has relieved his impatience. While still young, a bull can now sire as many calves in weeks as he would unaided in his whole lifetime, so that his daughters' yields can be compared with that of their mothers and his merit quickly assessed.

This 'progeny testing', judging parents by how their off-spring perform, is unquestionably the most accurate way of testing whether it is worth continuing to breed from these parents (and one ruminates on its human applications). It also gets us right away from what one writer tersely calls the 'Crufts approach', by which he means breeding for visible characteristics laid down in the rule books of the different Breed Societies, like inadequate noses in Pekingese or splayed front feet in Basset Hounds. Obviously, before the working of genetics was understood, it was sensible to judge by visible characteristics, but now cows, for instance, can be assessed by measurable attributes, such as milk yield and butter fat percentage. Pigs are submitted to an ultrasonic meter, which will accurately measure their back fat thickness and it is possible to construct an index to guide the breeder in his selection of parents. This is based on efficiency in converting food into flesh, economy in expensive feeds and the back fat over the eye muscle area.

The list of possible breeding aims in cattle, for instance, is formidably long: growth rate, efficiency in utilising food, adaptability to different environments, fertility, longevity, milk yield, the extent to which the carcase suits the butcher, and so on. A testing service for pedigree breeders was in fact set up in 1964, but nowadays, the stress is less on in-breeding or on breeding for a single factor, such as growth rate. We go more for compensatory breeding. An animal which is generally satisfactory is mated with one that is particularly good just where the first one is weakest. Many desirable characteristics are thus pooled and can then be fixed by subsequent in-breeding.

With milk production doubled in the last twenty-five years it is no surprise that the cowhouse of our farm, the pride and joy of its builder in 1849, long ago proved inadequate. What suited the man carrying two buckets on a yoke across his shoulders was obviously quite unsuited to the needs of the twentieth-century cowman and it is interesting that the main objections to the old cowhouse (apart from its obvious lack of hygiene by our standards) were its lack of light and of ventilation. So again new wine was poured into old bottles. The roof was raised; the floor concreted, with a dung channel running behind the stalls; the old wooden partitions torn out and replaced by metal tubing. But this proved money thrown on to the midden and quite soon the whole was torn down. It was replaced by a modern milking parlour and in-wintering quarters of the sort already described. Between this and the Dutch barn there is now a great silage tower, from which the silage is mechanically conveyed to the cows in their new quarters.

This is now a far larger milking herd than the original one and the acreage farmed from this steading has nearly trebled. The initials of the original builder and the date still appear in blue bricks against the red of one of the few surviving walls, but he would have difficulty in recognising the farmstead he so lovingly designed. However, the farm is no longer in the hands of his family, although country families are not easily uprooted and the name is borne by many people in various grades and walks of life in the nearby villages and market town. The farm itself is now owned by a large company and it is their local manager who lives in the farmhouse. Since 1849 in fact, the population of these isles has trebled. Nevertheless, as already proclaimed in chapter 4, in 1972 we produced all our milk, 60 per cent of our cereals and cheese, over 70 per cent of our animal feeds, over 80 per cent of all our meat and actually more pork, potatoes and eggs than we needed.

Today's manager would not think of having hens scratching around all over the place. He has not opted for the fully enclosed, 'nunnery' type hen unit, but away out of sight, two large Nissen huts each contain 1500 laying hens on deep litter. Although this is really too small an enterprise by modern standards, the straw is home grown; on it the birds wander freely and out of it they produce manure. Nor would they gain much if they were still loose, for the whole farmstead is now surrounded by a concrete apron. Mud has gone. In fact, a rotating brush, rather similar to those employed in street cleaning, can be run by the tractor's power take-off and used to tidy up when necessary. The whole area is lit up at night by powerful lights on the high gables.

A new building is now going up and it emphasises rather neatly the changes of the last few years. This rejects corrugated iron on steel girders in favour of pre-fabricated concrete supports, covered in asbestos-cement sheeting. The effect is therefore white, not black or red or green painted or even unpainted corrugated iron. It has the usual concrete floor, with vast sliding doors, and inside it is empty. A tractor can comfortably operate inside this new multi-purpose shed, which at one moment may contain a heap of grain, at another a pile of potatoes (for building them into clamps in the fields now demands far too much labour), or earlier in the year sacks of seed corn and piled up bags of fertiliser. Nothing could be more remote from the Victorian farmstead, with its beautifully laid, solid brickwork, its inflexible divisions and its narrow passages. The original buildings are indeed scarcely recognisable now. Yet, if you are instructed and observant, you will still be able to pick out little bits of the basic design of Victorian or Georgian farmsteads amongst the much larger and much less solid buildings of many a modern farm.

The naked electric light shines on the tractor in the shed before sunrise on a winter morning. Unexpectedly tall for its width, its canvas and Perspex cab conceals the roll-bar which is now a compulsory safety provision. (The modern high-powered machinery used on farms can be as dangerous as that in factories). Perspex flaps open on either side, but the wide hanging flap at the rear is the significant one, because through this the tractor driver will watch all day long whatever implement he is drawing behind him. And what you notice at once are the great rear wheels, quite disproportionate to the rest, emphasising that the prime function is to pull, just like the professional weight-lifter's exaggerated biceps. Again, in the evening the quiet is broken by the approaching roar of a tractor in top gear and behind it the bump and rattle of an empty trailer. As it turns off the road, the headlights pick out the old low wall of the cattle yard, before it turns into the shed and the driver switches off both engine and lights. Quiet returns and you can again hear the rustle of cattle in the yards. One more day's work done, another link in the long chain stretching right back through the generations to the time when the land was first cleared. For 'while the earth remaineth, seedtime and harvest . . . shall not cease'.

Index

For Product Safety Concerns and Information please contact our EU representative GPSR@taylorandfrancis.com Taylor & Francis Verlag GmbH, Kaufingerstraße 24, 80331 München, Germany

Printed and bound by CPI Group (UK) Ltd, Croydon, CR0 4YY

04/05/2025

01860342-0001